THE

ORIGINS

OF

FIELD THEORY

STUDIES IN
THE HISTORY OF SCIENCE

THE ORIGINS OF FIELD THEORY

L. PEARCE WILLIAMS

CORNELL UNIVERSITY

RANDOM HOUSE New York

To Sylvia

It is one of the boasts of modern science that it is a, if not *the,* truly open-ended intellectual system in which dissent is both welcomed and rewarded. The practitioner of science has been brought up on this idea and proudly repeats it until, perhaps, he finds himself on the side of dissent. Then the "open" ranks suddenly close and he finds himself isolated and alone, wondering how it happened that his careful adherence to the rules of the game has led to ostracism. Similarly, the layman is conditioned to accept the non-dogmatic nature of science and does so until he is confronted with the violent clash of two or more scientific systems. The educated resident of Berlin in the early 1920's, for example, must have been shocked by the epithets that were traded between the partisans and antagonists of relativity theory.

The open end of science seems frequently to belch forth fire and brimstone upon the unorthodox who dare to challenge the truths that science has so laboriously laid up for the contemplation and benefit of mankind. One way of resolving this seeming paradox of simultaneous dogmatism and ability to take up revolutionary new positions in science is to view science as it develops in historical perspective. Given the dimension of time, the paradox may be seen to consist of separate parts. The dogmatic element is always present for it represents the hard-won victories of the past. At any given moment in time, there is a body of scientific truths which most scientists would defend as being proven beyond doubt. Yet, time and time again it is the most dearly held truths which have to be sacrificed if progress in a given area is to be made. Nothing could be more obvious than the fact that an inanimate body does not move unless something pushes it; yet Galileo was to insist (for the wrong reasons) that bodies could move without a pusher—and Newton was to erect classical dynamics on this very point. From Newton's analysis, it followed that the velocity of a body must contribute to the velocity of any substance emitted from it—yet Einstein was to deny this "fact" and, thereby, revolutionize physics.

The purpose of the *Random House Studies in the History of Science* is to illuminate those frequent moments in the evolution of science when the scientific world finds itself torn between the old and the new. These are dramatic times, for men totally committed to an intellectual position are likely to defend this position with passion as well as with reason. It is this drama that the essays in the series attempt to recreate for the modern reader, both scientist and layman.

THIS VOLUME is the outgrowth of a biography of
Michael Faraday which I completed a few years ago.
Although Faraday was the major architect of classical
field theory, he also was a practicing chemist, teacher
and popular lecturer and these facets of his life had to
be treated if a picture of the whole man were to emerge.
Thus, the outlines of the development of field theory
tended to be blurred by these other aspects of his
career. The history of a concept has the advantage over
a biography in that it is possible to stick to one point
throughout and to follow the development of the ideas
(and only those ideas) which led to the mature concept. This is what I have tried to do in the present
volume.

The reader should be warned at the outset that much
of what follows is highly controversial in scholarly
circles. The influence of the German Nature Philosophers is by no means accepted by my peers; I can only
rest here on the fact that if their influence is admitted, a

great number of otherwise mysterious events in the history of early nineteenth century physics suddenly seem to make sense. My estimate of the relative worth of the contributions of Faraday and Maxwell to the development of field theory will also, I suspect, meet with opposition. Here my defense is somewhat stronger; I have only followed Maxwell's own estimate. For the documented basis of much which I say, the reader is referred to my *Michael Faraday, A Biography* (London, Chapman & Hall, Ltd., 1965 and New York, Basic Books, 1965).

I wish to take the opportunity to thank two of my colleagues with whom I have discussed my ideas and from whom I have received many insights. Professor Henry Guerlac's unsurpassed knowledge of Newton's *Opticks* and of the reception of Newtonian ideas in the eighteenth century was offered to me freely and was most gratefully accepted. Mr. Henry Steffens' excellent master's thesis (unpublished), *The Development of Newtonian Optics in England, 1738–1831*, provided the firm scholarly foundation upon which my discussion of the Newtonian optical tradition could be based. I would hope that my thanks here would repay both men in part for their help.

L. PEARCE WILLIAMS

Ithaca, New York
November, 1965

Contents

THE ORIGINS OF FIELD THEORY

CHAPTER

I

THE MATURITY
OF CLASSICAL
MECHANICS

IN 1687 there appeared in London a work which is probably the most important volume ever published in the history of science. Its title was severe—*Philosophiae Naturalis Principia Mathematica*—and its contents were forbidding. Throughout, the author, Isaac Newton, followed the rigorous road of geometrical proof, laying down theorems, propositions and lemmas in logical order. Behind the dispassionate presentation of the mathematical principles of natural philosophy lay a revolution in scientific thought. The product of this revolution was classical mechanics.

Newton alone among his contemporaries saw that there were mathematical principles of natural philosophy and that these principles could be generalized to include the whole universe of ponderable matter. Parts of this fundamental insight had been glimpsed earlier.

Johann Kepler had dimly perceived a mathematical harmony. His three laws of planetary motion had been presented almost casually to a somewhat uninterested world, whereas his belief in the mathematical harmonies of the ratios of the planetary orbits to one another had sent Kepler into ecstasies and convinced his more sober contemporaries that the *via mathematica* could lead only to folly. Galileo, on the other hand, although as convinced as Kepler of the essential mathematical nature of the cosmos, carefully refrained from universalizing his vision. To him, it was enough to enunciate the law of free fall and hope that the simple equation $S = \frac{1}{2} gt^2$ would serve as adequate evidence that God was a Geometer and had created the cosmos according to geometrical precepts.

Newton drew from both Galileo and Kepler. From the one (by way of Descartes) he took his first law— the principle of inertia. From the other, he assimilated the idea of force. His task, and it was a gigantic one, was to associate the two. Ponderable matter, all by itself, either did not move at all or, if moving, continued to move in a straight line. Force, which to Kepler had been a kind of spiritual emanation, to Newton became that which caused matter either to commence motion or to alter its path when in motion. Force acting on matter through time produced acceleration. Newton's first two laws of motion brought together the worlds of Kepler and Galileo.

The second law should be looked at from the back side, as it were. We are today accustomed to speaking of force producing an acceleration, for this is, after all, what $F = ma$ means. And when we apply force to a body by, say, pushing it, it does in fact accelerate. But what are we to do when there is no visible object around to push a body? This involves us now in the reverse application of the second law. If we see a body accelerating then we know that a force is acting upon it. How this force acts at a distance, as Newton was quick to point out in the *Principia,* may be veiled forever from our sight, but that it does act is indubitable. Nor should we be disheartened by our failure to penetrate behind the action to the cause of the acceleration, for there is something else about this force that we can discover: It is possible to find not only that a force acts but also the precise law which this action obeys. Thus, the force acting upon masses of matter which causes them all to deviate from the straight line motion demanded by the principle of inertia is found to depend upon the product of the masses involved and to vary inversely as the square of the distance separating their centers. The law of universal gravitation is thus given specific, mathematical formulation in spite of our total ignorance of what gravitation itself is. The entire universe could be summed up in Newton's three laws of motion and this law.

There was, however, a price to be paid for this simplicity. The real world of real bodies with all their richness of textures, colors, smells and sheer physical evidence of existence had to be left behind. In their place were mathematical points following mathematical curves through Euclidean space. Mass itself, that most obtrusive of all qualities, became an abstract noun. The mass of the earth or of the moon could be considered to be concentrated at a mathematical point at its center. It was this point which entered into the equations, not the earth itself. Thus, the mechanics that Newton founded contained within it certain implicit consequences which, in the eighteenth century, were to be avidly uncovered and accepted, particularly by Newton's French disciples.

Once the French mathematical physicists of the eighteenth century had been convinced of the errors of Descartes, they embraced the Newtonian system eagerly. Their embrace, however, was a peculiarly French one. They would have none of the agonizing of Newton and his compatriots over the ultimate nature of force or gravity and the possible religious consequences that might flow from the assumption of the inherent and essential presence of force in matter. If, in fact, the universe of point-masses, acting upon one another at a distance, could get along without the active intervention of the Deity, then *tant pis pour le bon Dieu!* He should, as the *philosophes* insisted, be content with having created a harmonious universe and that splendid creature, Man, whose mind could comprehend it.

Man's comprehension of the activity of ponderable matter was presented in terms of the mathematical equations describing its actions. Given point-masses and forces following the Newtonian definitions of their action, it was possible to reduce mechanics to a branch of mathematics. This was what Louis Lagrange accomplished in his *Analytical Mechanics,* published in 1788. Boasting that he had not used a single illustration or diagram, Lagrange proceeded to deduce the laws of mechanics in rigorous mathematical fashion from the Newtonian laws. The reality of ponderable matter was reduced to a series of differential equations.

The path opened up by the *Principia* was not the only product of Newton's genius. In 1704, he published a second scientific treatise which differed markedly from his first masterpiece. The *Opticks* was as wholly devoted to the experimental aspects of physics as the *Principia* had been to the mathematical.

In the *Opticks,* Newton revealed the purely physical side of his mind in which he no longer refused to "feign hypothesis." The dimensionless mass points of the *Principia* which had served primarily as the starting point of a differential equation were now given physical garb and reality. "It seems probable to me," Newton wrote in the 31st Query appended to the *Opticks,* "that God in the Beginning form'd matter in solid, massy, hard, impenetrable, moveable Particles, of such Sizes and Figures, and with such other Properties, and in such Proportion to Space, as most conduced to the End for which he form'd them; and that these primitive Particles being Solids, are incomparably harder than any porous Bodies compounded of them; even so very hard, as never to wear or break in pieces; no ordinary Power being able to divide what God himself made one in the first Creation." These atoms, it should be noted, performed more of a function than that required by the *Principia.* The "Sizes and Figures" which Newton assumed had been given them could be used to account for some of the qualitative aspects of physical phenomena. Could not color, for example, be assumed to be the result of the arrangement of these particles? And could not the chemical qualities of bodies likewise be reduced to the "Size and Figure" of the atoms? Such hints, thrown out by Newton, were to occupy the minds of the investigators of the eighteenth century. A number of very ingenious schemes were devised, all of which served to make atomic hypotheses commonplace. None of them, however, went beyond Newton's. They all offered possible qualitative explanations of events, but were unable to bring the atomic realm within the severe critical framework of experimental practice. Thus, the argument that acids corroded metals because the acid particles had sharp points and literally chiseled away the metallic particles from the metal surface provided a vivid picture of acid action but gave no clue as to why different acids acted in different ways on different substances. It was this failure to deal with specific reactions that led Antoine Laurent Lavoisier to write in the preface of his *Elementary Treatise of Chemistry* that speculations on the ultimate nature of the elements was a waste of time. Elements (and by extension, all chemical facts) were to be defined in terms of what could be determined in the laboratory, not by what could be dreamed up by this or that armchair philosopher.

It is against this background that the work of John Dalton must be viewed. The difference between Dalton's atomic theory and those of Democritus and Newton is a crucial one. Dalton brought atoms down from the ethereal realm of metaphysics to the cold reality of the chemical laboratory. All atoms of the same element, said Dalton, were identical and, most important, weighed exactly the same. What differentiated one element from another was the relative weight of the atoms of each element. This atomic weight was not merely a clever idea but an experimentally determinable fact.

For the first time, a physical parameter could be associated with the ultimate particles of matter. Atoms became the very center of chemical philosophy; the Newtonian skeleton had received a layer of muscle which was to reinforce belief in the system as a whole.

The successes of the *Principia* and its French commentators and of Dalton's atomic theory had been confined to the world of ponderable matter. To be sure, this was the world of ordinary and celestial mechanics, as well as of the chemical transformations which played such an important role in both natural philosophy and practical affairs. But alongside this ponderable world was another, more subtle one of seeming imponderables. The sun had been emitting heat and light in prodigious quantities for centuries without apparent loss of size. The conclusion seemed inescapable that light was either imponderable or of such small weight as to be negligible. Similarly, electricity and magnetism could be induced in bodies without any detectable gain in weight. The great question that confronted the eighteenth century was whether such phenomena could be brought within the framework of Newtonian science. Once again, it was Newton himself who led the way to show how light, at least, could be treated in roughly the same terms as the particles of ordinary matter.

Throughout most of the body of the *Opticks,* Newton preserved the same kind of non-hypothetical attitude that had marked his approach to mechanics in the *Principia.* Experimental demonstrations followed one another in close array as Newton drew the net closed around optical phenomena. Until the very end, Newton insisted upon dealing with the laws of optical action rather than with the nature of light. Whatever the nature of light may be, these laws were still valid and Newton could rest assured that his contribution to the knowledge of the action of light was a permanent one. Such a phenomenological attitude, however, is never entirely satisfying. One wants to know more than what happens; the *why* of an event cries out for an answer. Furthermore, as Newton knew full well, the suggestion of the *why,* in terms of a physical model, often permits the prediction of new effects about which the pure description of laws is mute. An hypothesis concerning the nature of light was therefore forthcoming but only in the Queries and not in the form of a flat statement. "Are not," Newton asked, "the Rays of Light very small Bodies emitted from shining Substances? For such Bodies will pass through uniform Mediums in right Lines without bending into the Shadow, which is the Nature of the Rays of Light." Light, like the elements of ponderable matter, was particulate in nature. To this extent, then, the worlds of ponderable matter and light were identical.

Did the identity extend beyond this? Or, did the particles of light interact with ponderable matter in ways essentially different from the laws of mechanics set out in the *Principia?* The answer, given in qualitative terms, could be drawn directly from Newton's experimental data. It was, simply, that ponderable matter and light, like ordinary bodies, acted upon one another at a distance.

Reflection, the most commonly observed optical effect, Newton showed could not be simply the result of the impact of the light corpuscles against the particles of the reflecting surface. A polished surface is, after all, merely smooth to the gross sense of human sight and touch. All one need do is look at a polished body with a microscope to see hills and valleys produced by the polish itself. If light corpuscles actually hit these wrinkles, they would be scattered, rather than reflected so as to form a coherent image. There must be, then, as Newton put it, "some power of the Body which is evenly diffused all over its Surface, and by which it acts upon the Ray without immediate Contact." This power was obviously one which repelled the light corpuscles. A similar power, but of opposite sign, could be utilized to explain refraction. As a light corpuscle approached a piece of glass, it would (depending upon the angle of incidence) first penetrate the repulsive power and meet an attractive power closer to the glass surface. This would then pull it into the glass; the attractive power acted in such a way that Snell's law was obeyed.

With these two forces acting at a distance upon the particles of light, Newton could account for all the really difficult problems of physical optics. The diffraction of light passing by a hair or other small object was obviously the result of the repulsive force pushing the light corpuscles out of their straight course. The phenomenon of "Newton's rings" was somewhat more difficult to explain. Why was it that if monochromatic light were used, a series of bright and dark rings appeared when a plano-convex lens was illuminated with its convex side resting upon a reflecting surface? Perhaps, Newton suggested, it was because the corpuscles of light ran into periodic disturbances of the ether through which they passed which prevented their passage through the surfaces of the lens. Such periodic disturbances were like shock waves, set up by the light corpuscles in the ether which, traveling faster than the corpuscles themselves, arrived at the interface before the corpuscles. When a corpuscle arrived at the interface, its reflection or refraction depended upon the state of the ether at that moment. If the shock wave had produced a momentary increase of density of the ether, the corpuscle was reflected; if, on the other hand, the ether was in the rarefied phase of the compressional wave, it would be refracted. To put it another way, a light corpuscle was repelled when approaching an area of higher ether density and attracted when approaching one of lower ether density than the one through which it was traveling. Here Newton combined a wave theory with the use of corpuscles. The "fits of easy reflection and refraction," as he called them, were periodic results dependent upon the waves set up in the ether.

There was one final property of light that Newton had to struggle with. When light passed through certain bodies such as iceland spar and other similar crystals, it suffered two different and simultaneous refractions. Could this be explained, Newton asked, in any other way than by assuming that light corpuscles, like little magnets, had sides or poles which were differently affected by the attractive and repulsive forces of the crystal? Thus, the two refracted rays were composed of light corpuscles whose poles were oriented in different and opposite directions; they were, in fact, polarized by their interactions with the crystalline forces.

This was the theory of light which Newton bequeathed to his successors. No one could deny that it was (and is) one of the great landmarks in the art of experimental reasoning. What should also be remarked, however, is the *ad hoc* character of the hypotheses Newton was forced to devise to give his theory completeness. Light must be composed of corpuscles because corpuscles travel in straight lines and the only other possibility—waves—had to be discarded since waves go around corners. But, of course, light does not travel in straight lines, as witness its diffraction by a hair. So, the corpuscles must be thrown out of their path by some force acting at a distance upon them. Light corpuscles, however, were not merely the passive subjects upon which external forces acted; they, too, were endowed with attractive and repulsive forces, concentrated at their opposite ends or poles. Nor was the interaction simply between light and ponderable matter. The hypothetical ether entered in and served to confuse matters by acting attractively or repulsively upon light depending upon whether it was in a rare or dense state. Were attraction and repulsion, then, really action at a distance and dependent upon the interacting particles alone, or were they functions of ether density and accomplished in some mysterious way by this quintessential medium? That was for Newton to know and the eighteenth century to find out.

There was no doubt that Newton's optics fascinated those who came after him. It served both as a guide to experimental science and as a stimulus to further investigation. On the theoretical side, there were two aspects that demanded further attention. There was the natural question of the relation between the *Principia* and the *Opticks*. Could not the laws of mechanics be applied to the laws of optics so that Newton's achievement could be seen as a single, harmonious whole? This task was to be accomplished primarily by two men, Robert Smith and Henry Brougham.

Just as important as this first task, and in a sense dependent upon its successful completion, was the necessity of turning Newton's queries into solid scientific knowledge. The corpuscular theory of light as Newton left it was almost entirely qualitative. Like the atomic theory before Dalton, it lacked precise physical parameters to flesh out its metaphysical core with solid physical facts. What Dalton had done for the atom, Brougham in Scotland and J.-B. Biot in France were to do (with less success) for the light corpuscles.

The job of applying the *Principia* to the *Opticks* was undertaken soon after Newton's death in 1727 by Robert Smith of Cambridge. In 1738, he published his *A Compleat System of Opticks in Four Books,* whose explicit purpose was to reduce the theory of optics devised by Newton to the mathematical principles of natural philosophy. Perhaps Smith's most significant contribution, for it was to recur throughout theories of matter, both ponderable and imponderable, in the eighteenth and nineteenth centuries, was his reconciliation of the simultaneous attractive and repulsive powers of matter for light. What had appeared in Newton as *ad hoc* hypotheses introduced to explain a particular effect (i.e., attraction to explain refraction and repulsion to explain diffraction) were now reduced to physical extensions of a general mathematical law. For, Smith wrote, "as in algebra where affirmative quantities vanish and cease there negative ones begin, so in mechanicks where attraction ceases there a repulsive virtue ought to succeed." The specific action at a distance (attraction or repulsion) was, therefore, a function of the distance across which the action took place. Furthermore, the attractive and repulsive forces clearly differed significantly from gravitational forces in that they did not extend throughout all space, but fell off rapidly with distance. Thus, Smith essentially defined two areas of mechanics: microphysics, in which one has to deal with intense, short-range forces which can be studied by the use of such probes as light corpuscles; and macrophysics, in which the primary force was that of universal gravitation. The assumption of the corpuscularity of light seemed essential to the exploration of the microphysical realm, and this in itself appeared to be a weighty reason for accepting it.

Yet, it still lacked substantiality. What was needed was a real physical parameter. This was what Henry Brougham provided in two papers which appeared in the *Philosophical Transactions of the Royal Society* in 1796 and 1797. Brougham's method was simplicity itself and reinforced Smith's use of simple mechanics to explain the inner mechanism of optics.

The deflections of rays of light as they passed near diffracting bodies could be measured with great accuracy. In this way, the law relating the falling off of the repulsive force with the distance from the diffracting body could be accurately determined. The speed of the light corpuscles themselves was fairly precisely known. From these two facts the relative sizes of the corpuscles making up the different color-fringes of a diffraction pattern could then be calculated. Thus, it could be shown that the color of a ray of light was uniquely determined by the size of its corpuscles, and the relative sizes of the different light corpuscles that gave rise to the colors of the spectrum could be calculated with great precision. Here, as later in chemistry, a real physical property had been discovered in the hypothetical light corpuscles. They, like the atom, had been brought by the new mechanics from the shadows of metaphysics into the bright light of physical existence.

The success of Smith and Brougham in reducing light to the simple action of corpuscles served to obscure the wave-properties which Newton himself had stressed. Neither Smith nor Brougham ever explained Newton's "fits" but merely explained them away. This was a serious weakness in the theory and could not be ignored. The French physicist J.-B. Biot provided the final element of the corpuscular theory of light by reducing the periodic aspects of Newton's "fits" to the oscillation of the light particles around a given axis. If, Biot argued, the particles of light are like little magnets with opposite poles, then they will be transmitted or reflected depending upon whether or not they present one or the other pole to the reflecting or refracting surface. The obvious periodic nature of the "fits" Biot attributed to the fact that passage through the first surface of a thin film or of a plano-convex lens caused the light corpuscles to oscillate about their polar axis. Whether or not a $+$ or a $-$ pole presented itself at the second surface, then, clearly depended upon the distance through which the particle had traveled. Hence, the "fits," as Newton had shown, would be functions of the thickness, and the periodicity was not due to waves but to oscillations of the particles. Unfortunately, the mechanism of these oscillations escaped Biot and was to prove one of the rocks upon which the corpuscular theory foundered.

In the first years of the nineteenth century, however, Biot's and Brougham's work seemed unassailable; optics was a branch of mechanics and could be treated by considering particles upon which forces of attraction and repulsion acted according to familiar and easily determinable laws. To challenge this was tantamount to challenging the whole structure of Newtonian mechanics.

The almost absolute faith that was put in physics at the beginning of the nineteenth century rested upon more than mechanics and optics. It had been strongly reinforced by the inclusion of electrostatic and magnetic phenomena within the Newtonian pale. Here, the triumph of Newton's partisans was particularly striking for they had to combat a fairly sophisticated theory of electric and magnetic action to which even Newton had shown some favor.

Before the publication of the *Principia* the prevalent concept of the mechanism of the transfer of motion from one body to another was that of impact. One body moved because another body pushed it, not because the other body attracted it across empty space. Thus, gravity, for example, as Descartes had pointed out, was the result of the impulsion given to bodies by the circulation of an extremely subtle matter about centers of gravitating force such as the earth and the sun. It was this theory that Newton had exploded in Book II of the *Principia* by showing that a circulating fluid simply could not do what Descartes demanded of it.

There were, however, certain phenomena in which "attraction" seemed palpably to be the result of a circulating fluid or effluvium. If, for example, iron filings were sprinkled upon a card and held over a magnet, the filings took up positions in curves that clearly revealed the path of the magnetic effluvium. Similarly, if a body were sufficiently charged electrically, the electrical effluvium could actually be felt as it issued from the charged body. Newton himself appeared to accept this model, for Query 22 of the *Opticks* asks "how an electrick Body can by Friction emit an exhalation so rare and subtile, and yet so potent, as by its Emission to cause no sensible Diminution of the weight of the electrick Body, and to be expanded through a Sphere, whose Diameter is above two Feet, and yet to be able to agitate and carry up Leaf Copper, or Leaf Gold, at the distance of above a Foot from the electrick Body? And how the Effluvia of a Magnet can be so rare and subtile, as to pass through a Plate of Glass without any Resistance or Diminution of their Force, and yet so potent as to turn a magnetick Needle beyond the Glass?" It was this line of thought, rather than that suggested by the *Principia,* which was generally pursued in the first half of the eighteenth century.

At first sight, the problem appeared to be relatively simple. The only real difficulty lay in the lack of resistance encountered by the effluvia, both electrical and magnetic, as they moved through other, more solid, bodies. Yet, even here, the way out was at hand. A necessary consequence of the corpuscular philosophy was that empty spaces must exist between the particles of ponderable matter. Indeed, the passage of the corpuscles of light through diaphanous bodies was impossible without the existence of such pores. All that was required was to adjust the relative sizes of the pores and the effluvium particles so that the one was enormously large compared to the other. Then, no resistance resulting from collisions between the atoms of ponderable matter and the effluvium particles could manifest itself. Immediately this obstacle was hurdled, however, another appeared which was born of the first. For, in point of fact, the theory requires that there be collisions, otherwise the observed "attractions" will not take place. The "attractions" are, after all, caused by the impulsions of the effluvia against the particles of the body being "attracted." If, for example, the electrical effluvium could pass through the pores of a piece of tissue paper without hitting any of the particles, then there would be no impulsion and, hence, no attraction. So, it was necessary to modify the theory and assign relative "pore-sizes" to different bodies. Gold, for example, must have very large pores compared to iron for it permits the magnetic effluvia to pass through it without effect, whereas iron is caught up in the current and swept to the magnetic pole. This hypothesis, however, is in conflict with a number of other facts. Gold certainly has a greater density than iron, and before Dalton density was viewed as the number of particles per unit volume with little or no attention being given to the weight of the individual particles. Hence, in this view, gold contained more particles packed into a unit volume than iron, yet the spaces between these particles must be larger to permit the free circulation of the magnetic effluvia—a clear and obvious contradiction. To get around this, an *ad hoc* hypothesis had to be forthcoming. The pores, it was suggested, were not just holes but shaped holes and the particles of the effluvia also had shapes. Some bodies, then, permitted the effluvia to enter; others did not. It was these latter which were swept along in the effluvial current. The model was getting a trifle clumsy.

As people wrestled with the theoretical difficulties associated with the effluvial theory, others turned to the intensive experimental study of the phenomena themselves, especially electrostatics, which was in its infancy when Newton wrote. Two discoveries in particular were to cause insurmountable difficulties to the effluvial theory. The first, reported by Stephen Gray, was that there appeared to be two kinds of matter—that which could be given an electrical charge (modern insulators), and that which could not (modern conductors). The second was that some electrified bodies *repelled* one another instead of attracting each other as the effluvial theory demanded. To work these facts into the theory required great ingenuity, and probably the finest (and last) effluvial theorist was the French Abbé Nollet. His comprehensive theory required no fewer than eighteen fundamental propositions by which all electrostatic phenomena could be explained. The first fourteen are reproduced here.

FUNDAMENTAL PROPOSITIONS, drawn from experience, & with the aid of which one can account for all the electrical phenomena known until now:

    1. Electricity is the effect of a fluid matter which moves around or within the electrised body.

    2. This fluid is neither the real matter of the electrised body, nor the coarse air which we breathe.

    3. There is every ground to believe that the electric matter is the same as that of elementary fire and light, united to some other substance which gives it an odor.

    4. This matter is present everywhere, in the interior of bodies as well as in the air which surrounds them.

5. The electric matter, excited or put in action, moves, as much as it can, in a right line, & its movement ordinarily is a progressive movement that transports its parts.

6. The electric matter is sufficiently subtle to penetrate through the hardest and most compact bodies.

7. But it does not penetrate them all with the same ease. Living bodies, the metals & water are those in which it passes most easily; sulphur, sealing wax, glass, resins, & silk are those in which it has the most difficulty penetrating, unless those bodies are rubbed or heated.

8. The air of our atmosphere is not as permeable to the electric matter as the metals, living bodies, water, etc.

9. When the electric matter leaves a body with great impetuosity and issues forth into the air, whether it be visible or not, it divides up into several divergent jets, which form a kind of spray or plume [brush].

10. A body electrised by friction or by communication emits, from all sides, rays of electric matter which extend in straight lines in the air or in the other bodies in the vicinity.

11. While these emanations continue, a like matter comes from all sides to the electrised body in the form of convergent rays.

12. These two currents of electric matter, which go in contrary directions, perform their movements at the same time; & one of the two is stronger than the other.

13. The pores by which the electrised matter leaves electrised bodies are not so numerous as those by which it re-enters them.

14. The matter which comes to the electrised body is not furnished to it by the air alone, but by all the other bodies in the vicinity which are capable of becoming electrised by communication.[1]

[1] I. Bernard Cohen, *Franklin and Newton* (Philadelphia: American Philosophical Society, 1956), pp. 388 ff.

With these and the remaining four propositions it is, in fact, possible to explain all the electrical phenomena known when thc Abbé Nollet wrote. Prop. 7, for example, explains the difference between conductors and insulators. Props. 10, 11, and 12 explain, equally well, the observed attractions and repulsions of charged bodies. When the affluent stream (that which proceeded away from the charged body) was stronger (Prop. 12), bodies were pushed away; when the effluent stream (that which proceeded from surrounding bodies to the charged one) was stronger, bodies were attracted. That such streams existed could be easily shown. One merely had to note the direction in which sparks leapt from or to a charged body. Could any intelligent person reject such a complete theory? The answer, even in the eighteenth century, was yes. And the reason for this rejection is quite instructive. It has little to do with "science" for Nollet's theory did explain the known facts quite adequately. But, could any theory that required eighteen *fundamental* propositions be the true one? Euclid, after all, had laid out the foundations of geometry in ten, and Newton had provided the basis for the science of mechanics in three. Must not the mechanics of the imponderables be as simple? What was needed, clearly, was a theory which could account for the facts without requiring such a complicated scaffolding of hypotheses. This is what Benjamin Franklin provided.

Franklin's attention was called to electricity by the performance of a "Dr. Spence" who toured the Colonies titillating his audiences with the latest electrical tricks. Franklin was fascinated! What was this strange power that appeared in rubbed glass and was capable of attracting chaff and straw to it? Fortunately for Franklin, he was largely unaware of the theories and hypotheses that European investigators had suggested. If he had read of the Abbé Nollet's system, it did not appear to have affected him unduly. The one scientific work which served as his guide was Newton's *Opticks* and this provided him with his theoretical framework. He did not adopt Newton's effluvial approach but, rather, the idea of action at a distance upon which Newton based his theory of light. Thus, Franklin's electrical theory bears a striking resemblance to Newton's ideas on the nature of light.

The matter of electricity was most probably even more subtle than that of light "since it can permeate common matter, even the densest metals, with such ease and freedom as not to receive any perceptible resistance." The particles of the electrical matter, however, were unlike those of either ponderable matter or light. "Electrical matter," Franklin pointed out, "differs from common matter in this, that the parts of the latter mutually attract, those of the former mutually repel each other." "But, though the particles of electrical matter do repel each other, they are strongly attracted by all other matter."[2] Ponderable matter, Franklin went on to say, was like a sponge, capable of absorbing a more or less great amount of the electrical fluid. When the sponge was squeezed so that electrical fluid was forced out of it, it then contained less than its normal amount and was negatively charged. When the

[2] *Benjamin Franklin's Experiments,* I. Bernard Cohen, ed. (Cambridge, Mass.: Harvard University Press, 1941), p. 213.

sponge had imbibed an excess of the electrical fluid, it was positively charged. The attraction of the electrical fluid for the particles of ponderable matter accounted for the observed reaction of straw and chaff to a charged body. The mutually repulsive action of the particles of the electrical fluid provided a theoretical basis for the distribution of charge on a charged metallic body as well as the repulsion of two positively charged bodies. Thus, Franklin's two hypotheses on the nature of the electrical fluid could do what it took Nollet eighteen fundamental propositions to accomplish. Furthermore, Franklin's theory was fully in accord with Newtonian mechanics. A generation had passed since "action at a distance" had first raised a storm. The mere repetition of the term had served to familiarize people with it. Familiarity may not always breed contempt, and it rarely generates heated opposition. Thus, Franklin's suggestion of applying action at a distance to electrical forces was greeted with cheers as a great simplification.

Only at one point was Franklin's theory vulnerable. It could not explain the repulsion of two negatively charged bodies, for why a mutual *lack* of electrical fluid should have this effect was a mystery. There were only two ways out of this difficulty. Either the particles of matter should be considered as possessing repulsive as well as attractive virtues, or *two* electrical fluids needed to be assumed. The first tack was taken by the physicist Franz Aepinus, in 1759. The second was suggested by Charles Coulomb in 1788. Both systems were to have their adherents in the nineteenth century.

Franklin's theory brought electricity no nearer to quantitative determination. His experiments, however, did. He was able to show, by a most ingenious and simple experiment, that the quantity of positive charge to be found on the inside of a charged Leyden jar was always exactly the same as the negative charge on the outside. The principle of the conservation of charge was thus forthcoming, but since the absolute quantity of electrical fluid still escaped measurement, the relations between negative and positive induced charges could not be simply calculated. The situation is roughly analogous to Newton being able to deduce Snell's law of refraction from his corpuscular theory.

The full quantification of electrostatics was achieved by Charles Coulomb, who also, thereby, strongly reinforced the fluid theory. By means of a torsion balance which he invented, Coulomb was able to measure very small forces with great accuracy. His first application of his new instrument was to the measurement of the viscosity of liquids, but the problem of electrical and magnetic attractions and repulsions soon caught his eye. In the late 1770's, the obvious similarity between electrical and magnetic action became so obtrusive that the Academy of Sciences of Bavaria offered a prize for the best essay on the subject. What emerged from this contest was a clear realization that before any definitive answer could be given to the question of whether or not electricity and magnetism were identical, the laws of their action would have to be precisely determined. This was just the kind of challenge Coulomb loved. He had a positive passion for measurement (while hiding out in the country from the forces of the Terror in the French Revolution, he passed his time by measuring the force of saps in trees), and measurement here obviously could solve the problem.

Coulomb's investigations led to results which must have gladdened the heart of every ardent Newtonian. The particles of electricity, Coulomb noted, acted *exactly* (and this exactly meant quantitatively) like the particles of ordinary matter. Newton's law of gravitational force between two particles of masses M_1 and M_2 could be expressed by the simple equation $F = \dfrac{kM_1M_2}{r^2}$, where k is a constant and r is the distance between the centers of M_1 and M_2. The attractive force between two "masses" of electricity Q_1 and Q_2 (Q_1 being of opposite sign from Q_2) followed the same law: $F = \dfrac{k_1Q_1Q_2}{r^2}$. Only k_1 differed from k. Two magnetic "masses" acted upon one another in the same way, again only the constant of proportionality being different. These equations gave a reality to the particles of the electric and magnetic fluids which they had not hitherto possessed. Furthermore, the differences in the constants of proportionality for electrical and magnetic action showed that the electric and magnetic fluids were not identical, but merely acted according to formally identical laws. By 1789, when Coulomb had completed his measurements, both electricity and magnetism could be understood completely in Newtonian terms. It is worth noting these terms explicitly.

There are two electrical fluids, one of which we call positive, the other negative. Particles of the electrical fluids are attracted by (and attract) the particles of ponderable matter. Particles of the same charge repel one another; particles of opposite charge attract one another. The electrical fluids move with greater or lesser facility *between* the particles of ponderable matter, much as water and other fluids penetrate the pores of sponges.

There are two magnetic fluids, the boreal and the austral (north and south). Particles of the same kind repel one another; particles of opposite kinds attract one another. Unlike the electrical fluids, the magnetic fluids do not circulate between the particles of ponderable matter. They are, rather, confined *within* the molecules of matter, each half of a molecule of iron, say, containing a certain amount of the austral fluid and the other half containing exactly the same amount of the boreal. Magnetization consists in the alignment of these dipolar molecules so that their similar poles point in the same direction.

Coulomb suspected that all molecules of ponderable matter contained the magnetic fluids but his attempts to detect this universal magnetism were unsuccessful. For substances such as iron and nickel, however, the theory seemed quite adequate.

During the same years in which Coulomb was bringing electrical and magnetic action within the area of Newtonian concepts, still another phenomenon was being reduced to the action of an imponderable fluid. In the seventeenth century, heat had almost universally been considered the result of the "intestine motion" of the particles of bodies. There was considerable evidence to support this view, for one could daily see cases of motion producing heat. The rubbing of an axle on its bearing, the hammering of iron or even the effervescence of strong chemical reactions were all "motions" which generated heat. To suggest that heat *was* motion seemed natural. In the *Opticks,* Newton expressly accepted this view.

The eighteenth century was to desert Newton here in favor of being true to a higher "Newtonian" idea. Electrical effects seem to have first suggested the material nature of "fire." References to the electrical "fire" were commonplace in the first half of the century and it is a small mental leap from the identification of the electrical "fire" with a unique substance and the extension of this to cover ordinary "fire."

Belief in the materiality of heat was reinforced by some peculiar behavior of snow noted by the Scotch chemist Joseph Black. While strolling in Glasgow one warm winter morning, Black was struck by the slowness with which the snow was melting. This was, upon reflection, a most odd thing! Why, Black asked himself, did not all the snow suddenly melt when the temperature rose over the freezing point of water? Instead, it appeared as though a great deal of heat was being absorbed simply to turn the snow into water. The laboratory confirmed Black's suspicions. A considerable amount of heat was needed simply to turn ice into water at 32° Fahrenheit. Similarly, the transformation of water into steam required the absorption of a considerable amount of heat without any increase in temperature. The heat that appeared to vanish beyond the reach of the thermometer Black dubbed "latent heat" for it could be recovered merely by reversing the direction of the change of state.

It is, of course, possible to account for latent heat by means of a kinetic theory but it is quite difficult. The sharp discontinuity between the liquid and the gaseous or the solid and the liquid state are not easily reconciled with the continuous increase in molecular motion that the addition of heat presupposed in the kinetic theory. Black was a chemist and it is not merely coincidence that the theory of heat which he favored was a chemical one. If heat were considered to be a material (though imponderable) fluid composed of separate particles, then the whole problem of latent heat could be easily solved. All that one needed to do was to make a few simple assumptions already honored in other areas. The particles of heat (christened "caloric" by Lavoisier) must be mutually repulsive since the effect of heat was to expand bodies. The particles of heat must also have an attraction for the particles of ponderable matter, for heat was always associated with matter. This association could be considered a fairly loose one, much like that of the electrical fluids with matter, and it was common at the end of the eighteenth century to think of material particles surrounded by an atmosphere of caloric. Temperature was a measure of the density of this atmosphere. Caloric could exist in a more intimate union, however, than this. It could "condense" out of the atmosphere and be bound tightly to the material particle. Such a union, akin in every respect to chemical combination, took the caloric particles out of the range of the thermometer and resulted in the formation of a new "compound" of heat and matter. Thus, change of state could be envisioned as a pure chemical reaction. Steam was not merely hot water in the state of a gas, but a combination of water and caloric. Furthermore, it was a *definite* compound whose proportions were as fixed as those of sodium chloride. John Dalton was to accept this view of heat and make it an integral part of his atomic theory.

By the beginning of the nineteenth century it was possible to see the combined effects of these separate developments we have considered. "Newtonian" mechanics had been transformed into a total world-view. Behind the bewildering complexity of appearances was a universe of utmost simplicity. There were ultimate particles of various kinds, but limited in their variety to a relatively few simple species. These particles acted upon one another across empty space or by bumping into one another. From these interactions came the infinity of observable phenomena. In the cases of some of these species—ponderable matter, electricity, magnetism—the laws of action and interaction were known. In the case of caloric and light, the laws were still somewhat obscured. One of the tasks of physics, then, was to seek them out.

For the theorist, the real problem of the future was merely a technical one. Pierre Simon de Laplace, one of the greatest of mathematical physicists, put the case in all its stark simplicity. The mathematical tools available to him (and to us) made it impossible to calculate the paths of more than two interacting particles. All one could do was to strive for ever more accurate approximations. Thus, the limit of man's knowledge was one imposed by his imperfect tools. Yet, said Laplace, if an infinite mind knew the position and momentum of every particle in the universe at any given instant, it could predict the future course of the universe to eternity. It was to this that Newton's work had led. Conceptually, the universe had been reduced to those elements which Newton had first introduced so cautiously in the *Principia*. There was no longer a need for new *ideas*. The next generation of physicists must rest content with searching for new techniques whereby the predictions of new particle locations and configurations could be made ever more accurate. If Laplace had only had a computer, then we might have seen something!

There is a tendency in every period for the brightest minds of that generation to feel that they have really exhausted all the ideas available on a given subject. So it was at the end of the eighteenth century. The "dogma" of action-at-a-distance physics was not one imposed by narrow-minded men who refused to see farther than their noses. It was, rather, a consensus held almost unconsciously by the best minds of the day. They would have protested vigorously against the term *dogma,* for this implies ramming a partial explanation down the throats of reluctant believers. Rather, they would have urged, this is simply the way physics is done; opposition to other ideas arose not from a fear of novelty but from a conviction of the essential correctness of the orthodox position.

One of the characteristics of orthodoxy seems to be that it stimulates heresy, and action-at-a-distance physics was no exception. Part of the heresy arose out of physics itself when the orthodox theory was found wanting. But, by far the greater part came out of a metaphysical revulsion for the pure mechanism of Laplacian physics. The heresy first appeared in Germany and then spread to England. Significantly, it never really penetrated France. Out of it was, ultimately, to be derived the fundamental tenets of classical field theory.

NATURPHILOSOPHIE AND THE DISCOVERY OF ELECTROMAGNETISM

NOT ALL THINKERS were satisfied with the Newtonian synthesis of the eighteenth century. To be sure, its grandeur and its seeming completeness appeared to leave little room for argument, yet there were still some who had nagging doubts. The most important area for criticism was also the largest; it involved nothing less than the power of the human mind to apprehend reality.

When we look at the Newtonian system as a whole from a philosophical point of view, one important point stands out: the philosophical jump from experience to the assumption of the existence of imponderable fluids by which observed phenomena could be explained. The followers of Newton in the eighteenth century attempted to erect their science upon the metaphysical principle that properties of things observable by our senses can be assumed to be the properties of ultimate particles which are inherently insensible. In its simplest form, this principle permits the reduction of various phenomena such as heat and light to the action of atoms or corpuscles having extension and figure moving through empty space. Newtonians thus accepted the ultimate reality of corpuscles or atoms and then loaded upon them the responsibility for accounting for the phenomena which are in fact observed.

This carrying over of sensible qualities to the insensible realm of atoms has recently been given the name of *transdiction*. The question that occurred to at least some philosophers in the eighteenth century was to what extent the process of transdiction is, in fact, legitimate. The most important such questioner of the very basis of Newtonian science was the German philosopher Immanuel Kant. It was Kant who in the 1780's first set out to discover the limits, if any, of the human reason. Essentially, he wished to know just how far the human reason could penetrate in its untangling both of the physical and of the metaphysical worlds. Three great critical treatises emerged—*The Critique of Pure Reason* (1781), *The Critique of Practical Reason* (1788), and *The Critique of Judgment* (1790). We shall be concerned here only briefly with *The Critique of Pure Reason* yet it is of fundamental importance for an understanding of the Kantian criticism of the underlying metaphysical assumptions of Newtonian science.

From the very beginning of *The Critique of Pure Reason,* Kant struck at the foundations of Newtonian science. For the Newtonian, time and space were absolute entities independent of the human mind and requiring only the existence of God Himself. Thus, events took place in absolute space, which could be conceived of as being completely empty. This preserved the necessary separation in Newtonian physics between the particles of matter, which were absolutely hard and impenetrable, and empty space devoid of all properties. Similarly, although in a somewhat more complicated fashion, time was considered to be independent of the human recognition of it. Processes took place in time as simple events without any other connection than their following upon one another. Causation, therefore, could be reduced (as it was by David Hume) to the mere sequence of events in the river of time. We shall see that Kant's critique of the absolute nature of space and time was to have a subtle effect upon this concept of causation which, in turn, was to be related to the basic concepts of field theory.

Kant subjected the Newtonian ideas of space and time to intense critical scrutiny. He simply asked himself how we could, in fact, know anything of pure empty space or of pure time in which no events occurred. The answer which he came up with was a somewhat startling one, namely that time and space do not exist independently of the human mind but are, in fact, the modes by which the human mind apprehends the existence of objects. Thus, *The Critique of Pure Reason,* if taken seriously as a fundamental critique of metaphysics, would mean the rejection out of hand of the very foundations of Newtonian physics. Particles could not act upon one another across empty space because empty space could not be known to the mind. Kant himself put this fact in no uncertain terms. In another treatise he wrote, "I do not mean to deny the existence of empty space: it may exist where perceptions cannot reach and thus no empirical knowledge of it can be gained; such a space is no possible object of our experience."[1]

The criticism of time as an absolute also had an effect upon the causal network of Newtonian science. After Hume, at least, the notion of a cause seemed to be restricted to a sequence of physical events which ultimately came down to the passing on of motion from one body to another, either by impact or by action at a distance. Thus, for example, when one wishes to know the reason or cause of the motions of the 8-ball in pool, it may be understood in terms of previous motion of the cue ball before striking it; if one knows the momentum of the cue ball then the exact path of the 8-ball can be deduced mathematically and the cause is merely the fact that A (the cue ball) has struck B (the 8-ball) and in some way passed on a motion contained in A. If, however, time is merely a mode of mental action or, to put it another way, the way in which the mind itself connects together phenomena, then more is required of a causal chain than mere sequence. What the Kantian looks for is not so much the fact that A with certain properties produced B with certain other properties by passing on some of these in a rather mysterious fashion, but rather the form of the entire process. Indeed,

---

[1] See Kant's discussion in Immanuel Kant, *Metaphysical Foundations of Natural Science*, E. B. Bax, trans. (London, 1883), pp. 150 ff.

process is the key word here for it involves a complicated series of mental activities for one to lay bare a
Kantian causal chain. We need to know not only the
sequence but the relations between the members of the
sequence, and there is a kind of architectural dimension
introduced by the Kantian criticism of absolute time.
We look for form and structure in processes and not
mere sequence of impact as in the corpuscular philosophy.

One very important consequence of this way of
regarding time and causation is the possibility of introducing interpretations into the observation of physical
fact. If, as the Newtonians would have it, physical
phenomena are merely events connected to one another
by a sequence in time, then the laws of the sequence
may come to be discovered and the world rendered
intelligible in terms of these laws. If, on the other hand,
time and space are the results of the synthetic activity
of the human mind, then the connections between the
events are connections that the human mind assumes to
see there. The fertility of these mental conceptions will
be determined by the manner in which these connections lead on to further discoveries and causal chains.
Thus precisely the same facts can be viewed in a
different causal connection to give insight into physical
phenomena. This will be shown to be precisely what
happened with the nature philosophers and the discovery of the correlation of physical forces—a correlation which to the Newtonian seemed improbable if not
impossible.

Kant's discussion of space and time in *The Critique
of Pure Reason* probably had little immediate influence

on the practicing scientist. The denial of absolute space with its concomitant questioning of the possibility of action at a distance across this space could be shrugged off as pure metaphysics. After all, a whole generation of scientists had been able to live with the mysteries of action at a distance and did not feel any great desire to resolve the mystery by throwing out the entire body of physics. The discussion of time was also, probably, received with indifference for the subtle distinction that Kant drew between time as an absolute and independent entity and time as the mode of perception of causality was one which most physicists probably felt was of little importance. After all, it didn't really make any difference as long as one had the proper laws of action which could be deduced from the Newtonian models of particles and action at a distance.

There was, however, one section of the *Critique* which did seem to have some direct relation to the prevalent mechanical philosophy. If all the laws of nature were ultimately to be deduced from the motions and forces of particles acting upon one another according to certain specific laws, it does become of some importance to know whether ultimate reality does consist of such particles. In short, was the atomic theory a legitimate theory, not only in terms of physical laws, but in terms of the human mind's ability to apprehend this ultimate reality? Kant's answer was a resounding no. He denied the fact that we could ever know, in any empirical way, the actual properties of those aspects of external reality which acted upon our senses. That there was such an external reality, he had no doubt, but that we could penetrate to its ultimate qualities he felt was impossible. Perhaps this point can be made clearer by a reference to our modern epistemological problem when dealing with fundamental particles. A high energy physicist may remark that a proton, when subjected to the action of high energy waves, becomes ellipsoidal, but one should not be led to believe that he really means that a spherical proton is changed by bombardment under high energies into a football; what he does mean is that if there were such a thing as very high energy waves which interacted with a proton whose shape was assumed to be a sphere, then, under certain conditions, the scattering pattern looks as if the spherical proton has been changed into a football-shaped proton. In fact, of course, one point of modern quantum mechanics is that we cannot represent the forms and properties of fundamental particles in ordinary language: the principle of transdiction simply breaks down. The physicist must therefore be content with highly sophisticated mathematical formulations which tell him nothing about the fundamental "physical" properties of his particle, if "physical" is understood to mean those properties which we recognize to exist in ordinary objects of our everyday experience. Similarly, Kant insisted that what our mind can know of matter has little relationship to what the matter is. The objects that exist as the basis of reality have certain effects on our senses. These effects should not be confused with reality itself. Thus if I feel a sharp stinging sensation when I thrust a needle into my tongue, I should not believe that the same sensation produced by dipping my tongue in an acid means that the particles of the acid have sharp points. This, to Kant, was an illegitimate transdiction and was therefore to be expelled from science. Yet, as we have seen, many of the explanations of Newtonian science were based to a great extent on precisely this kind of assumption: that properties of macrocosmic objects which produced specific sensations could be assigned to microcosmic objects without there being any possibility of serious error. If one accepted Kant's analysis, however, the whole mighty structure of Newtonian science fell to the ground.

Kant was not merely a critic of contemporary metaphysics but also a person intent upon constructing a metaphysics within which science could be rid of the epistemological errors that had crept into Newtonian science. It was not, however, in *The Critique of Pure Reason* that Kant sketched out his system of physics based upon his metaphysical ideas. This was done in another treatise which is, unfortunately, little read today but which is of fundamental importance for the history of nineteenth-century field theory. Its title is *Metaphysical Foundations of Natural Science,* and in this treatise Kant showed what legitimately could be used as the epistemological basis of a physics.

What, Kant asked, do we really mean when we speak of matter? Assuredly it is not little atoms bouncing around in empty space, for the distance between this idea and that of solid matter is almost infinite. What we actually mean, Kant argued, is essentially the resistance of an object to our attempts to move through it. That is, a table is not a congeries of atoms but a surface which pushes back on my hand when I push against it. This is matter, and the pushing back is the fundamental property by which I recognize that such matter exists.

Pushing back is, however, simply a force which repels my hand and prevents it from passing through the table without resistance. A repulsive force, then, seems clearly to be an essential part of our recognition of the existence of matter.

But repulsion is not all that is involved; the table, after all, does not swell to fill all space as it would do if it were merely a repulsive force in space. There is clearly some force which holds the table together as is clearly evidenced by any attempt to pull the table apart. The table resists such dissolution. Therefore, an attractive force must also be associated with the table if the table is to exist as an entity recognizable as a piece of matter of specific shape and size.

We have, then, two fundamental aspects of the existence of matter in these two forces. A repulsive force is necessary to prevent interpenetration; an attractive force is necessary to provide for the observed individuality of material objects. Kant would now argue that this is all we can know about matter; there are the two fundamental forces of attraction and repulsion associated with it, and the metaphysical foundation of any natural science must begin from this point.

Certain basic definitions then follow. In order to give the student some flavor of Kant's style and to allow him to realize why the Kantian system was so slow in penetrating the philosophical and scientific milieu of other countries than Germany, the following translation from the *Metaphysical Foundations* may be offered.

Proposition 1

Matter fills a space, not by its mere *existence* but by a *special moving force*.

DEMONSTRATION

The penetration into a space (in the moment of commencement this is called the endeavour to penetrate) is a motion. The resistance to motion is the cause of its diminution, and also its change into rest. Now nothing can be connected with any motion, as lessening or destroying it, but another motion of the same movable in the opposite direction. . . . Thus the resistance offered by a matter in the space which it fills, to all impressions of another [matter], is a cause of the motion of the latter in the opposite direction; but the cause of a motion is called moving force. Thus matter fills its space by moving force and not by its mere existence.

OBSERVATION

Lambert and others called the property of matter, by which it fills a space, *solidity* (a rather ambiguous expression), and maintained that we must assume it in everything *which exists* (substance), at least in the outer world of sense. According to their notions, the presence of something *real* in space must carry with it this resistance by its very conception; in other words according to the principle of contradiction; and must include the co-existence of anything else, in the space of its presence. But the principle of contradiction does not preclude any matter from advancing, in order to penetrate into a space in which another [matter] exists. Only when I attribute to that which occupies space, a power of repelling everything externally movable which approaches it, do I understand how it involves a contradiction, that in the space which a thing occupies, another [thing] of the same kind should penetrate. Here the mathematician has assumed something as a first datum of the construction of the conception of a matter, which itself does not admit of being further constructed. Now he can begin his construction of a conception from any datum he pleases, without committing himself again to the further explanation of this datum; but he is nevertheless not thereby permitted to explain the former as something wholly incapable of any mathematical construction, in order by this means to prevent a return to the first principles of natural science.

EXPLANATION II

Attractive force is that moving force whereby a matter may be the cause of the approach of others to itself (or, which is the same thing, whereby it opposes the retreat of others from itself).

Repulsive force is that whereby a matter can be the cause of repelling others from itself (or, which is the same thing, whereby it resists the approach of others to itself). The latter we shall also sometimes term *driving* and the former, *drawing* force.

NOTE

These are the only two forces of matter admitting of being conceived. . . .[2]

From this fundamental definition, which is in accord with sound epistemology in that all we can know of matter is its attractive and repulsive forces, certain physical principles in direct conflict with the atomic theory could be drawn. In Proposition 4, for example, Kant points out that matter is divisible into infinity into parts of which each is again matter. No more direct challenge to atomicity could conceivably be laid down.

For our purposes, however, there are two important concepts that should be stressed. In the first place, all reality is to be reduced to the two opposing forces of attraction and repulsion. The manifestations of these forces may occur in different ways, and, as we shall see, this was to become a point of extreme importance for

[2] *Ibid.*, p. 170.

the development of ideas on the activity and correlations of forces. Secondly, matter itself is defined as the filling of a space by force. Force or forces, therefore, are diffused throughout all space and there can therefore never be an empty space even if such were epistemologically possible. It would be saying too much to suggest that this is field theory, yet it contains the germ of field theory in so far as it focuses attention upon the space throughout which the forces are diffused. Kant is quite explicit on this point. He deals with light and with the forces of attraction to show that one cannot assume rays of particles carrying the force, as is the case with the corpuscular theory of light, but must rather think in terms of a continuum of force through which light and/or gravitation is propagated. The force of light, we know, is propagated at a finite velocity and it would seem legitimate to assume that other forces, even those of gravitation, have similar finite velocities of propagation. The finite propagation of force through a medium was ultimately to become one of the primary tenets of classical field theory.

Although Kant was relatively ignored in France and England during the 1780's and 1790's, he created nothing less than a philosophical revolution in Germany. Among his most ardent disciples was F. W. J. Schelling, who saw in the Kantian critique and reduction of reality to the forces of nature the possibility of creating a new philosophy more in accordance with his own nonmaterialistic, indeed mystical, inclinations. In a sense, what Schelling sought to found was a philosophy of mysticism in which experience itself could be brought to bear to buttress the visions of such mystics as Jacob Boehme and Johann Kepler in the sixteenth and early seventeenth centuries. This union of religious and scientific mysticism, Schelling felt, would be the best conceivable antidote to the materialism of Newtonian science and the social dissension brought about by this materialism in the French Revolution. It is important to recognize that Schelling's major goal was theological rather than scientific, but in his ascent from matter to God he left behind a number of ideas which were to have extraordinary influence in the development of physical theory at the beginning of the nineteenth century.

Schelling differed from Kant in one very important way. For Schelling the forces of nature were not, as they were for Kant, essentially dead and mere tensions existing in space. For Schelling the very essence of nature was the activity of the opposing forces of attraction and repulsion. Where, in short, Kant had the tendency to see a world in equilibrium between the opposing forces of attraction and repulsion, Schelling saw a world in conflict in which these forces constantly strove to overcome each other. Such a concept may seem silly to modern minds, yet it led Schelling on to certain important consequences. Perhaps the most important was the idea of polarity which such an interpretation of the opposition of attraction and repulsion leads to. An attractive force constantly battling with a repulsive force produced, in Schelling's view, a basic polarity in matter and, ultimately, in the whole universe. Thus it was that Schelling's universe was tied together by this polarity in a somewhat fantastic way. His fantasies deserve mention because they account in large part for the scorn that was heaped upon his followers, the philosophers of nature, in the generation that followed him. From the basic opposition of attraction and repulsion, Schelling rose to the basic opposition between light and darkness, which permitted him then to introduce the creative activity of the deity. "Let there be light," was the first commandment of polarity, and from this came the essential tension of the world both divine and human. Separation into sexes is a mere result of this polarity of the world. The lower organisms, which are essentially sexless, Schelling considered to be undeveloped, and his biological ideas were based largely on his principle of the evolution of the separation of male and female into separate individuals. Finally, of course, there was the duality of good and evil, which gave a theological cast to his whole system.

Such a continuity of polarity did, in fact, seem absurd to many but it also appealed to those who felt that there was a polarity in the universe which pervaded the moral, the theological, and material spheres. Certainly, in physics, it had some attraction since there existed recognized polarities that seemed inseparable. After all, positive electricity and negative electricity could be viewed as polar opposites. One never appeared without the other, and when positive electricity was opposed to negative electricity the two disappeared in a thunderclap of light and sound. Also, magnetism was a clear polar phenomenon; one kind of magnetism never occurred without its opposite and the two were inseparable. Polarity, then, seemed to be an important property of the physical world and this, too, was to have its effect in the development of field theory. We shall see Faraday wrestling with the problem of polarity and, at least in his early days, convinced that polarity was of such basic importance in electrostatic phenomena that he was willing to base his entire reputation upon its appearance in every electrostatic phenomenon.

One notion derived from the concept of polarity had a brilliant career in political and economic thought, even though it did not survive long in the history of science. This was the idea of the conflict of opposites leading not to an annihilation but, rather, to some new entity which in turn produced a polar opposite. The student of political theory will here recognize the famous Marxist dialectic: the conflict in history between thesis and its polar opposite, antithesis, leads not to annihilation of both, but to the creation of a new form or class, which in turn calls into being its own polar opposite. In this fashion Marx accounted for the dynamics of history, and the connection between Marx and Schelling and the nature philosophers is by now a clear one. We shall see later that this concept of activity in the uniting of polar opposites was to play an important role in the theoretical background of Oersted's discovery of electromagnetism.

Perhaps the most important contribution of Schelling to the development of field theory was his recognition of what was only implicit in Kant. If matter and material phenomena could be seen only as the results of attractive and repulsive forces, then it seems obvious that all phenomena should be reduced to these forces. Kant hinted at this but it was Schelling who really spelled it out. What Schelling tried to show in his *Sketch of a History of Nature* was that force is essentially dual. That is, that all the phenomena of nature can be reduced to the forces of attraction and repulsion. This basic duality and polarity underlies all the phenomena which are observed in the physical world.

Now what does this mean, if anything? It means that we must not turn to hypothetical entities to explain new phenomena. Or, to put it another way, the invention of imponderable fluids to account for optical, thermal, electrical, and magnetic phenomena, is illegitimate on a metaphysical basis. Instead, we have the manifestation of the basic forces (*die Grundkräfte*) under different circumstances as different phenomena. Electricity is merely the manifestation of attraction and repulsion under given physical conditions. Change the conditions and the manifestation changes. This is a point of such importance that it deserves to be underlined here quite explicitly. One of the traditional concepts of nature philosophy was the convertibility of one force into another. If the proper experimental conditions could be found, then it should be possible to turn each specific manifestation of force—light, heat, electricity, magnetism, gravitation—into another. Let it be noted that such a convertibility was not obvious in the Newtonian system: With separate fluids of distinct entities such as corpuscles of light, corpuscles of caloric, corpuscles of positive and negative electricity, corpuscles of north and south magnetism, it is not at all clear that one can transmute one kind of corpuscle into another. The convertibility of forces, which was to be one of the most important discoveries of the early nineteenth century, was an idea that was derived from nature philosophy and one to which the Newtonian system of physics was, if not hostile, at least indifferent.

So far we have talked philosophy, and many a reader

may be wondering what all this has to do with physics. Does it really matter if time, space, and matter are epistemologically impossible to conceive on Newtonian grounds or whether these entities are to be argued over by metaphysicians? What has all this to do with physics and the discovery of the laws of nature? What did a nature philosopher ever discover to make such a long discussion of nature philosophy legitimate in a book on the origin of field theory?

Fortunately, a nature philosopher discovered the most exciting new phenomenon of the early nineteenth century and he discovered it because he was a nature philosopher. I refer to Hans Christian Oersted and his discovery of electromagnetism in 1820.

Hans Christian Oersted was born in Denmark in 1777. His father was an apothecary, and his interests very early turned to the study of chemistry. Early in his life he learned to speak and read German and it was to Germany that he turned for his intellectual formation. The combination of German philosophy and chemistry was a natural one for Oersted and it was also an area in which a great deal of very exciting work was being done when Oersted went up to the university to study. As Schelling had pointed out, it was in chemistry that the real worth of nature philosophy would make itself felt, for here one had to deal with the essential points of the forces of nature. In chemical processes, after all, was to be found the origin of forms and qualities, the opposition of forces in chemical neutralization reactions, the evolution of light and heat in such reactions, and, in short, a microcosm of the universe as a whole. At the university, Oersted threw himself wholeheartedly into a study of German metaphysics and chemistry. His doctoral dissertation was written under the direct influence of Schelling's philosophy and aimed at the attempt to create a new chemistry according to the philosophical tenets of nature philosophy. In this period of his life was born his conviction of the unity of force, and it was this conviction which was to lead him on to the discovery of electromagnetism.

Oersted successfully defended his thesis in 1800 and almost immediately set out for a tour of Germany, where he was to meet some of the leading chemists and philosophers of the day. It was a particularly exciting time to be in Germany, for the air was filled with exciting ideas. In 1800, Alessandro Volta had announced to the Royal Society of London his discovery of a source of electrical current by which a continuous flow of the electrical fluid (in Volta's terminology) could be produced. Even before Volta's letter to the Royal Society was published in the *Philosophical Transactions of the Royal Society,* William Nicholson and Anthony Carlyle had shown that the passage of an electrical current through water caused the water to decompose into hydrogen and oxygen. Such a discovery was greeted with shouts of joy in Germany, for this, after all, was precisely what had been predicted by the nature philosophers. What it obviously meant was the identity of the electrical forces with those of chemical affinity, as Schelling had insisted. How the Newtonian could account for the decomposition of a chemical by the mere passage of another fluid through that chemical seemed to be a mystery, and the nature philosophers seized upon this new discovery as proof of the correctness of their views.

It was even claimed that this discovery invalidated

the new chemistry of Lavoisier. A number of ingenious experiments made in Germany seemed to show that the doctrine of the compound nature of water—a cornerstone of the new chemistry—appeared untenable in the light of experiments with voltaic electricity. Specifically, it seemed impossible to explain how hydrogen and oxygen, assumed to be the components of water and well known to be gases, could travel unseen through solutions to appear at opposite poles of the voltaic cell. It was with considerable glee that the German chemist and nature philosopher, J. W. Ritter, announced that the new electrochemical decomposition proved that water was a simple substance (i.e., an element) and that hydrogen and oxygen were merely water upon which the electrical forces had acted in different ways to turn them into gases. It might be noted in passing that there was more to this attack on Lavoisier's chemistry than pure science. Lavoisier, after all, had overthrown a chemical doctrine which had been created by the German G. E. Stahl, and 1800 was a date of particularly strong national feelings as Germany confronted the French Revolution and its threat to German culture and German national pride.

All this Oersted observed and heartily agreed with. He became the passionate partisan of Ritter's system, although he later recognized that Ritter was a careless experimenter whose speculations tended to outrun his experimental conclusions by far. Yet, the basic points of his compass remained true. He rested content with the tenets of nature philosophy and saw his task as the attempt to draw up a philosophy of nature in contrast to the Newtonian which would, in fact, free men's minds from the thralldom of a century of Newtonian orthodoxy.

Oersted's account of his attempt to create a rival system to Newtonian philosophy is to be found in an exceedingly rare book (only four copies are available in the United States), a work which has not been read with the attention it deserves. Published in Paris in 1813 with the French title *Recherches sur l'identité des forces chimiques et électriques* (*Researches on the Identity of Chemical and Electrical Forces*), it is far more than an attempt to establish, theoretically, what the voltaic cell had already seemed to prove experimentally. It concerned itself with the identity of all the forces of matter and, as such, was a direct challenge to the Newtonian view of the world. Oersted challenged the imponderable fluids as explanations of heat, light, electricity, and magnetism, and attempted, rather, to substitute the concepts of nature philosophy for those of Newtonian science.

As a chemist, Oersted's first concern was to bring the principles of nature philosophy into chemistry and show how they could clarify the problems which chemists faced in 1813. Chemistry was in a parlous state at this date. One of the important points that Lavoisier had made in his reform of chemistry was that chemical qualities were the result of the presence of certain chemical elements in a compound. This was stated most clearly in Lavoisier's identification of oxygen with acidic properties—indeed, the word "oxygen" means "acid former"—and Lavoisier had simply assumed that all acids contained oxygen and were acids because of this. In 1807 Humphry Davy had discovered the metals sodium and potassium, whose oxides formed the strongest alkalies then known. This discovery came as somewhat of a shock to the orthodox followers of Lavoisier, but they were nevertheless able to survive it by some agile mental gymnastics. In 1810, however, Davy showed to almost everyone's satisfaction that chlorine was an element and that hydrochloric acid was an acid which contained no oxygen whatsoever. What, then, was to be made of Lavoisier's doctrine of acids? All one could say was that an acid was a substance which turned litmus paper pink, and this is hardly a deep and satisfying explanation.

Oersted recognized that oxidation was a reaction of fundamental importance in chemistry and he therefore identified it with one of the fundamental forces of nature. As a good nature philosopher, he saw matter as composed of two fundamental forces in constant conflict, but the Kantian forces of attraction and repulsion were reduced by Oersted to the forces of combustion and combustibility. Every body contained one or the other of these forces in some kind of imbalance. If a body was primarily composed of the combustible force, then it burned in oxygen. If, on the other hand, a body was primarily composed of a combusting force, then the body permitted other substances to burn in it. These, for Oersted, were the two basic forces of nature from which he derived the idea of the forces of gravitation and of repulsion. This is the philosophy of Kant and Schelling as seen from a chemist's point of view.

It is of some interest to notice here the application by Oersted of the dialectic to chemistry: There are two essential classes of bodies in nature, combustibles and combusters, but these classes shade into one another, preserving that continuity which seemed to be of essential importance to the followers of Kant. But there is, nevertheless, a preponderance of one force over the other, so that we can think of substances as being essentially combustible or essentially combusting. The union of a combustible force with a combusting force is not annihilation but a new set of opposites. On the one hand, oxides form acids, as in nitric acid; on the other, they form alkalies, as in sodium oxide (or in its solution, sodium hydroxide), and we have again a union of opposites when an acid and an alkali combine to form a salt. This was the basis of Oersted's chemistry, in which the principles of nature philosophy are clear and which, in fact, did serve to call attention to certain chemical facts which were beyond the reach of the orthodox Newtonians.

Our real quarry here, however, is the unity of forces. We must now look at what Oersted considered matter itself to be and how matter was active by itself. An element, as we have seen, was a combination of the two opposing forces of combustibility and combustion. The equilibrium within the element was a relatively delicate one which could be disturbed by other forces impinging upon it. This is how Oersted explained the effect of electrical forces upon chemical compounds: The electrical forces produced by the voltaic pile were transmitted through the cell by a series of perturbations in the elements composing the pile. One should see this process as a kind of chemical vibration in which the electrical force temporarily destroyed the uneasy equilibrium between the two chemical forces; by thus destroying this equilibrium, the electrical force was itself transmitted through the voltaic cell and caused the decomposition of the compounds through which it passed. Electricity, then, would be defined by Oersted as a temporary disruption of the original chemical forces of an element or of a compound. It is of some importance to note this concept particularly. We do not have to deal here with fluids but with forces acting upon one another. The electrical forces are themselves produced by the chemical, as is clear from the action of the voltaic pile in which oxidations and deoxidations take place. Obviously, these chemical or electrical forces can be transmitted since they pass through a conductor to cause decompositions and recompositions in other chemical media. They should be viewed, however, merely as distortions of the original chemical forces, and what Oersted means by the identity of chemical and electrical forces is precisely this distortion that exists when a chemical reaction takes place. The distortion of the forces is transmissible from molecule to molecule, which are themselves but centers of patterns of force. An electrical current, then, may be viewed as a wave motion in which an equilibrium is temporarily disrupted and re-established and passed with great rapidity through a solid. Electrochemical decomposition is a result of the disruption and re-establishment of this equilibrium but with the further phenomenon of chemical separation and decomposition.

One factor that clearly emerges from Oersted's discussion of the identity of chemical and electrical forces is the ease with which the disruption of the original chemical equilibrium is transmitted. There is, in short, a gradation of conductibility and upon this gradation Oersted placed the greatest emphasis. When the disrupting force is easily transmitted through molecular distortions, the result is a simple electrical current. When this conduction of force meets with some resistance, the result is chemical decomposition and an electrical current. When the resistance to the passage of the disrupting force is stronger, there is the development of heat since heat, in Oersted's view, is merely the conflict of electrical force with chemical force in its passage. It should again be noted that the nature philosophers were among the first to accept the experiments of Rumford on the nonmateriality of heat, although they also rejected his idea of a kinetic theory in which heat was to be considered merely as the motion of the constituent molecules of bodies. When the conduction was increasingly hampered, light was to be found, and light itself was merely the transformation of the disruptive force into a constrained force that gave the sensation of light. Thus the electrical force, depending solely upon the circumstances of its condition through matter, could assume all the guises of the forces of matter.

What then of magnetism? Was not magnetism also to be found among the theories of the forces of matter? The answer, I think, should be given in Oersted's own words, for as early as 1813 he envisioned the possibility of the transformation of electricity into magnetism. It was then that he wrote: "The form of galvanic activity is the mean between the magnetic form and the electrical form." By this Oersted meant that the production of electricity through chemical action in the galvanic cell lay between the production of magnetism and the production of static electricity in an ordinary electrostatic generator.

"The forces [in galvanic activity] are more latent

than in electricity and less so than in magnetism." In short, the galvanic forces were somehow more deeply related to the basic forces of matter than were those of static electricity but, as he wrote on the next page, "it will be necessary to see if electricity in its most latent state does not have some action on a magnet as such. This experiment will not be without difficulty because the electrical action will always tend to mix in with it and make observation very complicated." In 1813, Oersted had envisaged the transformation of the electrical force into a magnetic force. It was to take seven years for this experiment to be realized. An account of the actual discovery can be given in Oersted's own words, for in 1830 he wrote an article for an encyclopedia in which he described his discovery of electromagnetism as though it had happened to someone else.

Electromagnetism itself, was discovered in the year 1820, by Professor *Hans Christian Oersted*, of the University of Copenhagen. Throughout his literary career, he adhered to the opinion, that the magnetic effects are produced by the same powers as electrical. He was not so much led to this, by the reason commonly alleged for this opinion, as by the philosophical principle, that all phenomena are produced by the same original power. In a treatise upon the chemical law of nature, published in Germany in 1812, under the title *Ansichten der chemischen Naturgesetz*, and translated into French, under the title of *Recherches sur l'identité des forces électriques et chymiques*,[3] 1813, he endeavoured to establish a general chemical theory, in harmony with his principle. In this work, he proved that not only chemical affinities, but also heat and light are

---

[3] Oersted here misquotes the title of his work. See above, p. 51. The reader should be told that the somewhat bizarre turns of phrase in this passage are Oersted's.

produced by the same two powers, which probably might be only two different forms of one primordial power. He stated also, that the magnetical effects were produced by the same powers; but he was well aware, that nothing in the whole work was less satisfactory, than the reasons he alleged for this. His researches upon this subject, were still fruitless, until the year 1820. In the winter of 1819–20, he delivered a course of lectures upon electricity, galvanism, and magnetism, before an audience that had been previously acquainted with the principles of natural philosophy. In composing the lecture, in which he was to treat of the analogy between magnetism and electricity, he conjectured, that if it were possible to produce any magnetical effect by electricity, this could not be in the direction of the current, since this had been so often tried in vain, but that it must be produced by a lateral action. This was strictly connected with his other ideas; for he did not consider the transmission of electricity through a conductor as an uniform stream, but as a succession of interruptions and re-establishments of equilibrium, in such a manner, that the electrical powers in the current were not in quiet equilibrium, but in a state of continual conflict. As the luminous and heating effect of the electrical current, goes out in all directions from a conductor, which transmits a great quantity of electricity; so he thought it possible that magnetical effects could likewise eradiate. The observations above recorded, of magnetical effects produced by lightning, in steel-needles not immediately struck, confirmed him in his opinion. He was nevertheless far from expecting a great magnetical effect of the galvanical pile; and still he supposed that a power, sufficient to make the conducting wire glowing, might be required. The plan of the first experiment was, to make the current of a little galvanic trough apparatus, commonly used in his lectures, pass through a very thin platina wire, which was placed over a compass covered with glass. The preparations for the experiments were made, but some accident having

hindered him from trying it before the lecture, he intended
to defer it to another opportunity; yet during the lecture,
the probability of its success appeared stronger, so that he
made the first experiment in the presence of the audience.
The magnetical needle, though included in a box, was disturbed; but as the effect was very feeble, and must, before its law was discovered, seem very irregular, the experiment made no strong impression on the audience. It
may appear strange, that the discoverer made no further
experiments upon the subject during three months; he
himself finds it difficult enough to conceive it; but the extreme feebleness and seeming confusion of the phenomena
in the first experiment, the remembrance of the numerous
errors committed upon this subject by earlier philosophers,
and particularly by his friend *Ritter,* the claim such a
matter has to be treated with earnest attention, may have
determined him to delay his researches to a more con-
venient time. In the month of July 1820, he again re-
sumed the experiment, making use of a much more considerable galvanical apparatus. The success was now evident, yet effects were still feeble in the first repetitions of
the experiment, because he employed only very thin
wires, supposing that the magnetical effect would not take
place, when heat and light were not produced by the
galvanical current; but he soon found that conductors of
a greater diameter give much more effect; and he then
discovered, by continued experiments during a few days,
the fundamental law of electromagnetism, viz. *that the
magnetical effect of the electrical current has a circular
motion round it.*[4]

In July 1820, Oersted sent a short communication in
Latin to all the learned societies of Europe announcing
the discovery of electromagnetism. The particular terms

[4] *The Edinburgh Encyclopaedia* (Edinburgh, 1830), re-
printed in H. C. Oersted, *Scientific Papers,* 3 vols. (Copenhagen, 1922), II, pp. 356 ff.

of this announcement reveal both the dimensions of Oersted's discovery and the confusion of his ideas in announcing them. After describing the effect, Oersted concluded:

> We may now make a few observations towards explaining these phenomena.
>
> The electric conflict acts only on the magnetic particles of matter. All non-magnetic bodies appear penetrable by the electric conflict, while magnetic bodies, or rather their magnetic particles, resist the passage of this conflict. Hence they can be moved by the impetus of the contending powers.
>
> It is sufficiently evident from the preceding facts that the electric conflict is not confined to the conductor, but dispersed pretty widely in the circumjacent space.
>
> From the preceding facts we may likewise collect that this conflict performs circles; for without this condition, it seems impossible that the one part of the uniting wire, when placed below the magnetic pole, should drive it towards the east, and when placed above it towards the west; for it is the nature of a circle that the motions in opposite parts should have an opposite direction. Besides, a motion in circles, joined with a progressive motion, according to the length of the conductor, ought to form a conchoidal or spiral line; but this, unless I am mistaken, contributes nothing to explain the phenomena hitherto observed.[5]

Thus it was that nature philosophy had led Oersted to a discovery which has immortalized his name in the history of science. It should be insisted upon that it was nature philosophy that was responsible for the discovery, since the orthodox physicists of the day simply

[5] "Experiments on the Effect of a Current of Electricity on the Magnetic Needle, by John Christian Oersted . . . ," *Annals of Philosophy,* 16 (1820), p. 276.

did not believe in the possibility of the conversion of forces in which Oersted had such faith. One can perhaps see this by referring to a letter that André-Marie Ampère wrote to a friend in 1820:

You certainly have a right to ask why it is inconceivable that no one tried the action of the voltaic pile on a magnet for twenty years. However, I believe that the cause of this is easily discovered: it simply existed in Coulomb's hypothesis on the nature of magnetic action; everyone believed this hypothesis as though it were a fact; it simply discarded every possibility of the action between electricity and so-called magnetic wires; the restriction was such that when M. Arago spoke of these new phenomena [of electromagnetism] at the Institute his remarks were rejected just as the ideas of stones falling from the heavens were rejected when M. Pictet read a memoir to the Institute on these stones. Everyone had decided that all this was impossible. It was the same thing that prevented the admission of the identity of the electrical and the magnetic fluids and the existence of electrical currents in the terrestrial globe and in magnets just as in the years gone by people refused to believe in chlorine as a simple body. Everyone resists changing ideas to which he is accustomed. It is amusing to see the efforts that certain people make to bring the new facts into accord with the gratuitous hypothesis of two different magnetic fluids and two electrical fluids simply because they are accustomed to think that way.[6]

There could be no doubt of the importance of Oersted's discovery, nor of its relation to nature philosophy as a whole. Certainly the assumptions of this school of thought seemed, finally, to have been justified by the

[6] L. de Launay, ed., *Corréspondance du Grand Ampère*, 3 vols. (Paris, 1936–43), II, p. 566.

new discovery. Yet, there was much in Oersted which remained to be explained. It is well and good to speak of forces and their transmission, their conflict, their metamorphoses. But what does all this really mean? How is a force transmitted? How do forces come into conflict with one another? What is the circular conflict around the wire which Oersted describes as the magnetic effects of the conflict of the electrical forces? These were not simply questions of terminology but of concept. They were not clear; they were not self-evident, and it required much more work to prove that the vision of reality that Oersted had used to guide him in his researches was, in fact, a vision that led to the truth or that could be comprehended by all men, even those to whom the German language was strange if not bizarre. Let us close this chapter with a résumé of Oersted's leading thoughts and, by making them specific and precise, show how it was possible to proceed from them to further discoveries on the unity and convertibility of forces.

We have already seen that the central idea of Oersted's system was that of the conflict of forces. How this conflict existed escaped Oersted and must also escape others for he was never to clarify it, nor were future researches to make it at all obvious. We must simply assume that it meant something to Oersted and let it go at that. But there is another idea contained within Oersted's work that is of particular importance for an understanding of field theory, namely, the idea of the conflict of forces spread out through infinity and covering all space. Oersted's attempt to produce magnetism from electricity was based on this idea of a web of force in which various strands played the part of the observable forces of nature. The universe as a whole could be seen as a three-dimensional net in which forces crossed one another in conflict or in harmony. When one force was stimulated by outside activity, its own vibrations gave rise to other forces. All space was filled with these forces; their manifestation as light, heat, electricity or magnetism depended upon the conditions under which the conflict took place. When confined to very small volumes, the conflict was known as chemical affinity. When the conflict could spread with relative ease across space it became the conduction of the electrical forces. When the conflict was restricted to a linear flow it was known as heat; when this conflict became too constricted for the easy conduction of the conflicting forces the result was light, and as Oersted's account of his own discovery shows, he felt that if this linear conduction could be still more straitly contained the result would be magnetism. Here are the germs of field theory, developed from the hint that was left by Kant. All space now is filled and crisscrossed with forces which manifest themselves according to the various conditions that exist locally in any state. Such a plenum of forces is the first step, at least, to the realization of a field theory in which the energy of any action lies in the continuous medium surrounding bodies rather than in the action of the bodies themselves.

There were also some other difficulties involved in Oersted's concepts. Although he attempted, as a chemist, to create a field chemistry, his arguments leave one singularly unsatisfied for he could not really explain the fine structure of chemical phenomena. To be sure, he could explain the various reactions that took place between combustibles and supporters of combustion, acids and alkalies and so on, but why specific reactions go with specific energies and why others do not go at all was not made clear by Oersted's discussion of the two basic forces of chemistry. Furthermore, the specific forms of chemical compounds—i.e., their crystalline arrangements and their specific identities—could not be deduced from the mere conflict of the forces that he suggested. What was needed, clearly, was a system in which such fine structure could be accounted for in terms of the forces of matter, and this was not to be forthcoming in Germany but in England. It was through the work of Humphry Davy and, particularly, that of his disciple, Michael Faraday, that field theory was to progress from an explanation of chemical fact to a universal system of the world.

CHAPTER

III

THE ELECTROSTATIC
LINE OF FORCE

OERSTED's discovery electrified the scientific world. Upon receipt of the news, every scientist in Europe and America hastened to his laboratory to repeat Oersted's experiment and to stare in wonder at an effect that had escaped everyone for the twenty years following Volta's discovery of current electricity. There must have been many who silently kicked themselves for not having seen what now appeared so obvious.

If Oersted's discovery astonished his contemporaries, his "explanation" of the origin of electromagnetism merely embarrassed them. No one flatly came out and said that the helical "conflict" of forces that Oersted seemed to see so clearly around a current-carrying wire was a hopelessly confused idea—Oersted had, after all, been led to his discovery by this view. But, the scientific journals of Europe received an ever-increasing stream of articles in which the conversion of electrical force into magnetic force was examined with minute care and its cause eagerly sought. By 1821, indeed, the problem appeared solved, for in the year following Oersted's discovery, André-Marie Ampère accomplished what, in July 1820, had seemed impossible. It was Ampère who reduced the swarming and conflicting *forces* of *Natur-philosophie* to Newtonian order. The imponderable fluids which Oersted had triumphantly thrown out the window now marched victoriously back into the temple of science through the front door.

Ampère had shared the astonishment of all when he first heard of the discovery of electromagnetism. With that speed of thought that characterized his mind, Ampère immediately leaped to a conclusion that appeared eminently worthy of test. If one current-carrying wire was accompanied by magnetic force, should not the magnetic forces of *two* wires interact? The experiment was not a difficult one to perform and Ampère was gratified by its success. Two wires carrying electric currents traveling in the same direction attracted one another; when the currents moved in opposite directions, they repelled one another. Tentatively Ampère drew a revolutionary conclusion: magnetism was merely electricity in motion. Again, it was a conclusion that permitted experimental test and Ampère set furiously to the task. If electricity in motion was magnetism, then it should be possible to reproduce all the effects of permanent magnets by suitably arranged current-carrying wires. Again Ampère was successful. A wire wound into a helix had all the properties of a magnet. Permanent magnets, then, were simply iron bars in which electrical currents flowed in circular currents perpendicular to the axis of the bar. This was Ampère's first suggestion, although why such currents

should circulate around the axis of a magnet was something of a puzzle.

It was Ampère's friend Augustin Fresnel, the pro-
ponent of the wave theory of light, who made Ampère's
hypothesis untenable by pointing out to him in a private
letter that permanent magnets revealed no sign of the
presence of electrical currents in them; they were not
noticeably hotter than their surroundings, although iron
was not a good electrical conductor. Nor did a magnet
decompose water when plunged into a glass of it. No,
co-axial electrical currents would simply not do. To
soften the impact of his destruction of Ampère's theory, Fresnel made a suggestion which was to prove
most fruitful. Since all that was required were electrical
currents whose *resultant* effect would be a co-axial current, why not simply assume electrical currents around
each iron molecule and then envision the process of
magnetization as the alignment of these molecules? In
this way all difficulties could be avoided for, as Fresnel
put it, we know absolutely nothing of the laws regulat-
ing the behavior of molecules and their environment so
that electrical currents *might* circulate around molecules and not be expected to have the side effects
accompanying electrical currents on the macroscopic
scale.

Thus was born Ampère's famous electrodynamic
molecule. It needed only one further element to be
complete. Coulomb's arguments on the necessity of two
electrical fluids still seemed valid to Ampère, so he
defined an electrical current as the flow of positive and
negative electricity past one another in opposite directions. In the molecule, the currents originated deep in

the molecule's interior and then flowed outside the molecule in opposite directions from one pole to the other, thus giving an effect equivalent to the flow of one fluid around the whole molecule. What must have been Oersted's chagrin to find that the electrical fluids, far from being discarded from science, were now to be an integral part of the theory of matter itself!

Ampère's electrodynamic molecule was only the starting point, not the end point, of his investigations. Armed with this physical model, he was able to undertake the mathematical analysis of the new science, which he christened *electrodynamics*. As memoir after brilliant memoir flowed from his pen, his contemporaries gradually lost sight of the physical cumbersomeness of his model and had nothing but admiration for the brilliance and elegance of his mathematics. The reality of the electrodynamic molecule rested upon such an imposing structure of mathematics that only a mathematical illiterate could doubt of its existence.

Michael Faraday was a mathematical illiterate. He was also a chemist to whom Ampère's whirling currents were, at best, an encumbrance and, at worst, an impassable barrier to the understanding of chemical processes. Finally, he had been touched by the currents of *Naturphilosophie* and had no fondness for theories which depended upon imponderable fluids flowing hither, thither and yon. Faraday, who proved to be Ampère's most serious opponent in the first half of the nineteenth century, was the father of field theory. In large part, the early concepts of the field grew out of his decade-long dialogue with Ampère. It will repay us later if we here examine Faraday's early ideas on the nature of matter and force, for they played central roles in the development of field theory.

In 1813 Faraday, then a young man of twenty-two years, was hired as a laboratory assistant at the Royal Institution of Great Britain. The man he was to assist was Sir Humphry Davy, then at the height of his fame. By 1813, Davy had made all the most important discoveries he was ever to make, with the exception of the miner's safety lamp. In the process he had had to wrestle with theoretical questions of the first magnitude. Unlike his contemporaries, he was not satisfied with the simplicity of Daltonian atomism, for he was too sensitive to the finer structure of chemical processes. How, for example, could the Daltonian explain the fact that some substances react with one another while others do not if the only force associated with atoms was that of universal gravitation whose embrace was indiscriminate? And how did the electrical "fluids" bind together substances into compounds? That these "fluids" were the chemical cement had been amply shown by Davy's own researches in electrochemistry.

Such questions appeared unanswerable in terms of the orthodox chemical theories of the day. Davy's dissatisfaction with these theories was enhanced by the influence of his good friend, Samuel Taylor Coleridge, the poet. In 1798, Coleridge had visited Germany and had returned to England with a strong case of *Naturphilosophie*. During his lifetime, Coleridge was noted for the brilliance of his conversation, and it was undoubtedly through this means that he infected Davy. In the early 1800's there was a striking change in Davy's terminology. Of a sudden he dropped all reference to imponderable fluids and spoke again and again of "energies," "powers," and "agencies." Although he never became the enthusiastic advocate of *Naturphilosophie* that Coleridge was, he appears to have suffered a quiet conversion to its tenets and attempted, albeit with great caution, to do creative chemistry within its framework.

As a chemist, Davy was quick to appreciate the fundamental weakness of *Naturphilosophie* as a general explanation of chemical phenomena. Opposed to the doctrine of the convertibility of forces, with which Davy could wholeheartedly agree, was the *fact* of the inconvertibility of substances. If all reality were, as Oersted so strongly urged, merely the result of the antagonism of opposing forces, why could not the conditions of this antagonism be changed in such a fashion as to bring about the conversion of substances (transmutation) as well as the conversion of forces? Or, to put it another way, chemical individuality was a brute fact that no amount of metaphysical argument could sweep away. If *Naturphilosophie* were to aid chemistry, then some means must be found of reconciling the universal antagonism of opposing forces with the specific individuality of chemical substances.

Fortunately for Davy and Faraday, there was a hypothesis available to them that did just this. It was an atomic theory which dispensed with matter, utilizing only the forces of attraction and repulsion, and thereby eminently adaptable to the epistemological requirements of *Naturphilosophie*.

The theory had been suggested in the eighteenth century by a Jesuit priest, Roger Joseph Boscovich.

Boscovich was a Newtonian who soon discovered that there were some hidden difficulties in Newtonian atomism. If, as Newton had categorically stated, atoms were perfectly hard and indivisible (i.e., without internal parts) then the description of the impact of two atoms becomes very difficult. Elastic collisions in the macroscopic world, as between two billiard balls, are made possible by the fact that the internal parts of the billiard balls can move relative to one another. Thus when one billiard ball hits another, the molecules of each are pushed out of place, the force of cohesion pulls them back into place, and it is this deformation and recovery which leads to elastic rebound. Such a description cannot apply to atoms since they have no parts to be displaced. Thus, Boscovich argued, atomic impact and rebound must be instantaneous, which means that at the moment of impact each atom has two simultaneous velocities in opposite directions, which is absurd. Boscovich saw only one way around this absurdity and this was to eliminate the "absolutely hard, impenetrable" core of the Newtonian atom. It was to be replaced by attractive and repulsive forces acting in such a way that all the properties of Newtonian atoms could be preserved. To illustrate this, we must have recourse to a diagram.

In Figure 1, the y-axis represents the intensity of the forces involved. Above the x-axis is repulsion, below, attraction. The x-axis is a measure of the distance from the origin, which is the center of the atom and is represented by a mathematical point. If we now bring in a detector particle from the right it will follow the hyperbola of universal attraction until it reaches H. The character of this attractive force will then change as the particle leaves the macrocosmic world, decreasing to zero at D and, as the particle continues to approach the

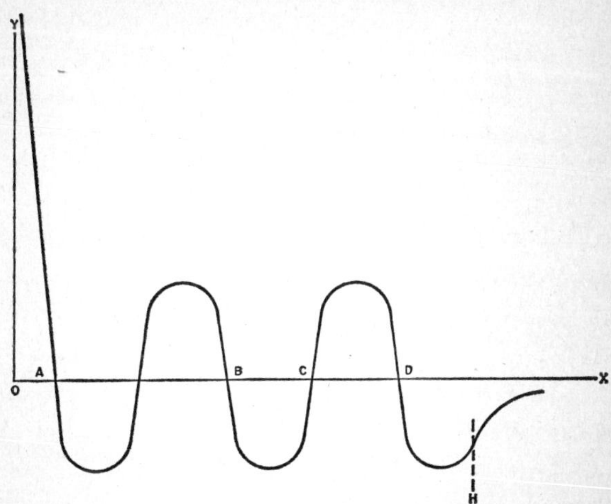

Figure 1

atomic center, turning into a repulsive force which increases and decreases in the hump between C and D. At C, there is another change of force, and so on, for as many "humps" as one needs to explain phenomena. At A, the character of the force undergoes another fundamental change. As the distance to the origin decreases, the repulsive force increases in such a way that it is infinite at O, thus preserving the essential property of impenetrability.

If one forgets the problem of why forces should act this way, there is much to recommend this atomic model. First, it solves the problem of atomic impact since the curves of force of two colliding atoms are continuous and therefore the process of impact and rebound will also be continuous. Second, and more important from the chemical point of view, the combination of groups of these atoms in stable configurations will give rise to particular patterns of force which are characteristic of specific groups. Thus, it might be argued, chemical individuality is the result of the formation of stable groups of force-atoms, and the process of chemical combination may be clearly envisaged as the interaction of various patterns of force. It is not at all unlikely that certain patterns will not mesh and therefore that two "elements" will not react with one another. Finally, since chemical affinity was simply the interaction of patterns of force, the distortion of these patterns might be expected to affect the affinities of substances. If electricity were considered to be such a distorting force, then the phenomena of electrochemistry might very well be rather easily explicable in these terms.

It was this theory which Humphry Davy found ever more attractive as he wrestled with the results of his experiments. And, at Davy's side, taking all in with eagerness, was his assistant, Michael Faraday.

Faraday's early interests did not lie particularly with electricity and magnetism. His earliest papers were strictly confined to analytical chemistry and the chemistry of chlorine, which particularly fascinated him. To be sure, Oersted's discovery excited him, as it did everyone else, but it did not touch off any important train of researches on electromagnetism. Not until 1821 did Faraday turn his attention to this subject. His friend Richard Phillips, editor of the *Annals of Philosophy,* then asked him if he would undertake a historical survey of the experiments and theories of electromagnetism which had appeared in the year following Oersted's announcement and pronounce upon their validity. This Faraday agreed to do but, as must happen when a brilliant mind comes into contact with a controversial series of facts and ideas, the result was not merely a critical survey but a most important discovery.

On September 3, 1821, the first really peculiar electromagnetic phenomenon caught Faraday's eye. Using a simple compass needle to follow the magnetic force around a current-carrying wire, it suddenly occurred to Faraday that this force was not only circular, as Oersted had dimly seen, but that an isolated magnetic pole should travel around the wire in circles. Faraday immediately constructed an apparatus which illustrated this effect; a bar magnet was suspended by one pole from a wire in such a way that the bottom end could rotate freely. The other pole was plunged into a glass filled with mercury in which a current-carrying wire was fixed upright. The whole formed a circuit and when the current passed, the free end of the magnet turned around the fixed wire, following the circular line of force. Faraday did not here use the term "line of force," but this was the moment of its birth. In following the evolution of the concept of the line of force in Faraday's mind we will be led to the origins of field theory.

Upon the publication of Faraday's paper describing the new electromagnetic motions, an interesting dialogue took place between Ampère and Faraday. What was at stake was nothing less than the conceptual basis of Newtonian mechanics. It is also the first time that Faraday went on record as being in opposition to Newtonian mechanics.

Ampère objected strongly to Faraday's use of the term "circular force." Newtonian mechanics was based on the existence of only one kind of force—central forces—which acted *in straight lines* between the origins of the forces. Thus the gravitational force acting between bodies A and B acts directly along the line connecting the centers of A and B. Similarly, it was assumed that the electrostatic or magnetic forces acted directly in a straight line between the charged particles or magnetic poles. The concept of the line of force is here of little use since it is a straight line whose origin and terminus are always known.

What, then, was Faraday's "circular force"? It must be, Ampère argued, simply the resultant of all the central forces generated by the electrical current. For Ampère, then, the *primary* fact was the poles or current elements in the wire from which the electromagnetic forces radiated; these forces, when summed vectorially, produced the circular force. An important element in Ampère's ardent support of this position was that such a system could be handled fairly easily with the calculus, whereas Faraday's insistence upon the primacy of the circular line of force led to great mathematical difficulties.

Faraday was unimpressed by Ampère's arguments. In the first place, as an experimentalist, Faraday restricted his belief in reality to what he could produce in the laboratory. Ampère's current elements and central forces, as far as Faraday was concerned, were entirely hypothetical. That they permitted Ampère to do some rather abstruse mathematical calculations meant very little to Faraday, who could not follow them. And Ampère's insistence upon the fact that the circular force was the resultant of central forces stimulated Faraday to perform a piece of turnabout. In a simple experiment, he showed that central forces could be the resultant of the circular force. This, in turn, led him to an experiment which convinced him of the reality of the line of force.

To show that central forces could be simulated by his circular force, Faraday simply took a straight wire and bent it into a circle. Now, all the circular forces within the circle were squeezed together and separated outside the circle. The intensification of the forces on each side of the plane of the circle of the wire led to a condition which was exactly similar to that of magnetic poles. If a pole of a permanent magnet was brought near this circle, it reacted just as though it had been brought near another magnetic pole. If the pole of the permanent magnet was taken to the other side of the wire, its reaction was as though it had been brought near a magnetic pole of the opposite sign. The effect could be made more evident by placing a number of such circles together, thus causing a separation of the "poles." This merely meant winding a wire into a helix and, as Ampère had already shown, a helix did display all the characteristics of a magnet. Thus, Faraday argued, the supposedly polar (or central) forces of magnetic action were really only the result of the concerted action of a large number of the circular lines of force surrounding a current-carrying wire.

To prove that magnetic attractions were not the

result of interaction between two magnetic poles, Faraday performed another ingenious experiment. He wound a helix around a glass tube and then suspended it half-submerged in a pan of water. He then took a magnetic needle as long as the helix and fixed it to a cork so that it would float. When the current was turned on in the helix, the north pole of the floating needle moved toward the south pole of the helix. Now, argued Faraday, if this were a simple case of the north pole of the needle attracting the south pole of the magnet, when the two met, they should cling together with some force. But, this did not happen! The needle entered the helix, and the north pole of the needle continued to traverse the tube until it reached the *north* pole of the helix. There it finally came to rest.

The conclusion Faraday drew from this experiment was of some importance. The curves formed by iron filings sprinkled on a card over a magnet were lines of force which a single magnetic pole (if such existed) would follow around and *through* the magnet, for the motion of the needle clearly showed that the line continued in the magnet and did not terminate at the poles. These curves, which were greatly to influence the course of his researches, now began to assume a reality in Faraday's mind.

Their very form was sufficient to give him pause. Such continuous curves seemed difficult to reconcile with simple action at a distance. Was it not more sensible to conceive of these curves as lines of *strain* in the surrounding medium? One of the fundamental characteristics of point-atoms was their capacity for undergoing precisely the kinds of strain that a magnet seemed to impose on them. After all, all point-atoms were linked together by their forces, and disturbance of this linkage constituted a strain which was propagated through space from one Boscovichean molecule to another.

Faraday's clearest use of this concept of strain in his early researches occurred in his discussion of the induction of electrical currents by a moving magnet. In 1831, he had discovered that a magnet could induce an electrical current in a wire when the wire was made to cut the magnetic lines of force. Now the question was how much current was induced and did this amount vary according to the material of which the wire was composed. Faraday's concept of the line of force as a line of intermolecular strain permitted him to deduce a law which very few of his contemporaries could see at all clearly. The line of force was a line of Boscovichean molecules to which a certain amount of tension had been applied. The amount depended solely upon the strength of the magnet. The case is exactly analogous to that of a rope being stretched by two opposing forces. A strain of 20, 40, or 60 pounds can be put upon it, but the strain is specific and is the same everywhere in the rope. When the rope is cut, exactly this amount of strain is released. It makes no difference what kind of knife is used to cut the rope; the strain is in the rope, not the knife. If this analogy is now applied to the line of force, a conclusion of far-reaching importance can be drawn. In modern terms, the electromotive force in a line of force depends solely upon the strength of the magnet producing it. Thus, the total e.m.f. will be directly proportional to the number of lines cut, regardless of the material of the wire which is cutting them. This being the case, then the *current* produced will depend only on the facility of the cutting wire for conducting this force. This current will then depend directly on the electromotive force and inversely upon the resistance of the wire. In 1832, Faraday clearly stated the same law which Ohm had enunciated in 1827 but which had remained in almost total obscurity since then. Faraday was led to it by his concept of the line of force as a line of intermolecular strain.

The further development of Faraday's ideas on intermolecular strain suddenly shifted in 1832 from the area of magnetism to that of electrochemistry and electrostatics. It was not so planned by Faraday, but came about as a result of some unexpected results of experiments which so intrigued him that he abandoned his magnetic researches to pursue them.

The occasion for this change of direction was the question raised in 1832 on the identity of all the electricities that were then known. Was the electricity produced by the electric eel, the voltaic pile, the electrostatic generator, moving magnets and the contact of dissimilar metals at different temperatures really all the same thing? Faraday was convinced they were and had, indeed, planned a whole series of researches on the basis of this conviction. It was not enough, however, to have a personal feeling of this identity; it must be proved. This Faraday set out to do.

The most stubborn fact that stood in the way of accepting the identity of the electricities was that static electricity simply would not decompose water, whereas this was the most striking activity of voltaic electricity. Prior to 1832, the attempt had been made to decompose water or salt solutions by discharging a spark from an electrostatic generator through a solution. The results were never clear-cut because of the physical disturbance of the solution by the violence of the discharge itself. With characteristic ingenuity, Faraday removed the disturbing causes by passing the electrostatic discharge through a wet string, thereby slowing down its passage and permitting it to act more like a current. The solution to be decomposed was then spread out on a glass plate so that the volume effects could be minimized. Now, when the electrostatic discharge was passed through it, there was true electrochemical decomposition and there could be no doubts of the identity of static and voltaic electricity.

At this point, most experimenters would have written up their results and gone on to other things. This was never Faraday's way. Having found an effect, it was necessary to look at it from every possible angle both to make certain that the observed effect was a real one and not just caused by accidental circumstances and for the simple pleasure of playing with a new phenomenon. One of the changes Faraday rang on his basic experiment can be seen in Figure 2.

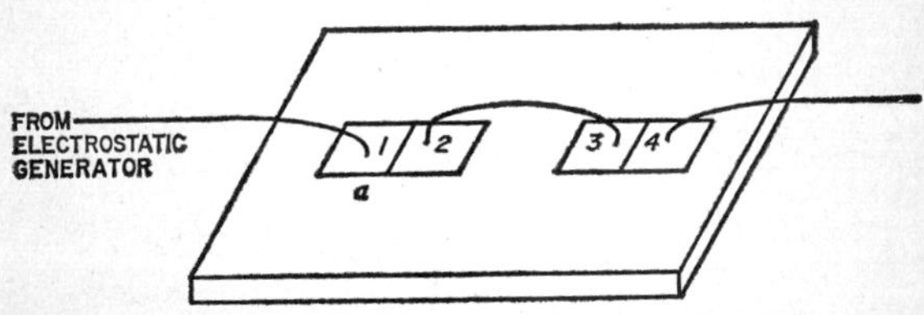

Figure 2

In this set-up, there were really no poles as such. There was merely a path along which the electrostatic discharge could pass; still electrochemical decomposition took place. Now this was a very astonishing thing! According to the accepted theory, the poles of the voltaic cell were the centers of attractive and repulsive forces which literally tore the molecules of a solution into pieces. The action was presumed to be action at a distance, the fragments migrating to their respective poles under the influence of the attractive and repulsive forces. One of the triumphs of early-nineteenth-century science had been to reduce the puzzling phenomena of electrochemical decomposition to such neat Newtonian terms. Now Faraday appeared to upset this particular part of the Newtonian world-view.

Faraday was never one to leap hastily to a conclusion. Before he was willing to take the risk of denying electrochemical action at a distance, he had to assure himself that, in fact, such action did not take place. There were two points at which he felt he could attack with some hopes of success. The first was the question of poles, for if poles were eliminated, as such, they could not serve as centers of force. The second was the existence of what we would today call free radicals in the decomposing solution. In dilute solutions, the "fragments" of the molecules that were being decomposed ought to exist in a free state for some time and they should then be detectable by ordinary chemical means. The reader should be reminded here that the concept of an ion as an electrically charged particle with peculiar chemical properties of its own was some fifty years in the future.

Faraday's experiment to prove the nonexistence of poles was simplicity itself. Instead of a circuit, he set up a discharge train so that the electricity merely passed into the air as a spark. A long strip of blotting paper was shaped so that it was pointed at one end, thus facilitating the discharge into the air. It was then soaked in a salt solution and connected by the blunt end to the electrostatic generator. Discharge of the generator through the blotting paper into the air led to the decomposition of the salt, even though there was no pole whatsoever to be found. Faraday's own words are worth citing here:

> Hence it would seem that it is not a mere repulsion of the alkali and attraction of the acid by the positive pole, etc. etc. etc., but that as the current of electricity passes, whether by metallic poles or not, the elementary particles arrange themselves and that the alkali goes as far as it can with the current in one direction and the acid in the other. The metallic poles used appear to be mere terminations of the decomposable substance.
>
> The effects of decomposition would seem rather to depend upon a relief of the chemical affinity in one direction and an exaltation of it in the other, rather than to direct attractions and repulsions from the poles.[1]

So much for the poles. We shall return soon to Faraday's concept of the relief and exaltation of chemical affinity.

The problem of the free "fragments" was a more difficult one to solve. As the above citation from Faraday shows, he was now quite skeptical about action emanating from the poles and this, in turn, made the likelihood of "fragments" really being produced rather remote. Yet, Faraday wished to test this conclusion as

---

[1] *Faraday's Diary*, Thomas Martin, ed., 7 vols. (London, 1931–36), II, ¶ 103–4.

fairly as possible. The failure to detect the free fragments in years gone by had been explained either by citing the density of the solution, which made recombination of the fragments an almost instantaneous thing, or by insisting upon the rapidity with which the fragments moved through the solution, making their detection extremely difficult. If Faraday were to prove that these fragments did not really exist, he would have, then, to eliminate both these objections. The first was easily done; he prepared a dilute solution of sodium sulfate. Into this solution, he then introduced some ordinary gelatine and made a sodium sulfate gelatine disc. On each side of this disc he placed indicator paper, then covered both these papers with two gelatine discs made with distilled water. On the outside of these discs he placed two more pieces of moistened indicator paper, and then pressed the two electrodes, in the form of plates, from a voltaic battery against these end papers. Thus, if there were free fragments produced by the action of the poles, they would be formed in the center slice and would have to pass through the pure gelatine discs, where, surely, their presence would be detected by the indicator paper separating the sodium sulfate jelly from the water jelly. When the current was turned on, however, these papers remained white, and only those at the electrodes turned color. As Faraday had written before, "the metallic poles used appear to be mere terminations of the decomposable substance."

If the poles were not able to tear molecules apart, then how did electrochemical decomposition take place? In Faraday's phrase, decomposition "would seem rather to depend upon a relief of the chemical affinity in one direction and an exaltation of it in the other." This phrase can be understood only in terms of Boscovichean molecules, where the pattern of forces represents the forces of chemical affinity. What the electrodes effected in this situation was a distortion of this pattern of force in the molecules contiguous to them. This distortion was then passed on to neighboring molecules in such a way that the distortion caused the constituent parts of each molecule to move past each other to join with another partner coming from a neighboring molecule. Let us assume, for simplicity's sake, that a molecule of water consists of one particle of oxygen and one of hydrogen, bound together by a mutually intense peak of attractive force. This molecule is surrounded by billions of other similar water molecules and, it should never be forgotten, all force-molecules are in mutual contact since their forces extend to infinity. Now, when the electrical force is applied at the electrodes, the area of intense attractive force in the oxygen particle is displaced in one direction, and the area of intense attractive force in the hydrogen particle is displaced in another. This displacement occurs in *all* the molecules and the result is that the oxygen particle from the original water molecule finds itself attached to the hydrogen particle of its neighbor, say, to the right, while the hydrogen particle of the original water molecule moves over to its oxygen neighbor to the left. The result is a series of chains of molecules in which the constituent parts move past one another in opposite directions. At no time, however, are the "fragments" free; they are always bound to a partner even though the partners change constantly.

Two aspects of this theory deserve particular mention. There is, first, the implicit notion of a threshold voltage required for electrochemical decomposition. Up to a certain point of strain, there should be no migration at all since the disturbance of affinity must be sufficient to carry over to a neighboring molecule. Up to this point, too, we have to do simply with an electrochemical line of force. It was the failure to discover any such threshold voltage that led later to the refutation of Faraday's theory.

The second point is of fundamental importance for the genesis of Faraday's electrical field theory. It was simply that electrical force in the case of electrochemistry was transmitted from molecule to molecule, not at a distance across empty space. It was a small enough beginning but it was the first really serious challenge to Newtonian physics that Faraday had made. The physicists, naturally, paid no attention to it, for who wanted to become involved in the messy smells and conditions of electrochemistry? It was not long, however, before Faraday threw the gauntlet squarely on the ground before them for, in 1838, he challenged the whole structure of electrostatic theory, built with such loving experimental and theoretical care by the French mathematical physicists.

The transition in Faraday's mind from the study of electrochemical to electrostatic forces came about naturally. The voltaic cell produced electricity by chemical means and Faraday asked himself what the situation must be like in a chemical cell before an electrical circuit was made.

> Is not rubbed glass and the rubber [he wrote] exactly in the state of Zinc and the oxygen of water in an electromotive circle, i.e. when the rubbed glass and the

rubber are separated are they not in the state assumed by the zinc and the oxygen before they combine and before the contact is made in a single voltaic circle. They probably give an exalted view of the conditions of the particles of the zinc and oxygen, a permanent view as it were. How do the states agree?

Would not this view, if supported, reduce both modes of evolution to one common principle. The mutual influence of neighbouring particles—in the glass not proceeding to a full effect; in the voltaic circle being completed and being followed in succession by a multitude of others of the same kind. In the last it is the attraction of the Zinc for the oxygen of the oxide, and this would tell in well for the instances of induction, etc. perhaps, of common electricity.[2]

From the beginning of his electrostatic researches Faraday was thinking heresies. If, as he suspected, electrostatic induction were similar to the beginning of electrochemical decomposition, then it must follow that the electrostatic force, like the electrochemical, was transmitted from particle to particle and did not act at a distance. Furthermore, if the transmission of the electrostatic force depended upon the intervening particles of the medium through which the force was passed, the character of these particles might be assumed to have some effect upon the force itself.

Faraday set out first to test this latter conclusion. He constructed an apparatus consisting of a brass ball which could be charged, surrounded by a much larger one which could be connected to an electrometer. The space between could be filled with diverse materials and the resultant induction through them be measured from

<hr>

2 *Ibid.*, ¶ 1646–7.

the outer sphere. As Faraday had suspected, a given charge on the inner sphere produced quite different inductive effects upon the outer sphere when different substances were placed in the hollow space. Furthermore, each substance had a specific inductive capacity which was characteristic of that body. This should not come as a surprise, for it follows directly from the theory of Boscovichean molecules. The molecules of every substance have a characteristic force pattern, and since it is this pattern which is involved in the transmission of the electrostatic force, we should expect that there would be specific differences between substances, akin to the differences in chemical affinity.

Faraday's discovery of specific inductive capacity was of the greatest importance. Its implications deserve to be examined with some care.

For Faraday, perhaps the salient point was the confirmation of his hypothesis that electrostatic action was intermolecular and not at a distance. Equally important was the fact that the discovery of specific inductive capacity forced a serious revision of Coulomb's law of electrostatic action. It was no longer enough simply to write $F = \dfrac{kQ_1Q_2}{r^2}$. It now became necessary to specify that this was in a homogeneous medium and also to specify what this medium was. Thus, if a two-inch sheet of sulfur were placed between the inductor and the body whose degree of electrostatic induction was being measured, the measured F would change without any change of distance. Also, Coulomb's constant, k, which he had assumed to be merely a constant of proportionality as universal as the gravitational constant, was really the specific inductive capacity and must be changed when electrostatic induction took place through different substances.

Finally, and most important for the future of physical theory was the extension of the field concept to electrostatics. As in electrochemical decomposition, the energy of the process is to be found in the medium between the electrostatic charges. The inducting body and the body receiving the induced charge are merely the termini of the line of electrostatically strained particles, just as the electrodes of an electrochemical cell are the termini for the lines of decomposing particles. The electrical *force* is to be found in the particles, not the termini. This is Faraday's electrical field theory. Note the change this forces upon Newtonian physics. Previously, one needed to know only the position and momentum of bodies to determine their future positions. The forces acting upon them were assumed simply to act at a distance. Now one also had to ask what the medium was in which these bodies existed for this affected the forces acting upon them. The space between bodies had previously been measured merely by a mathematical line; it now became a physical entity to be ignored only with great risk of inaccuracy.

While the existence of a specific inductive capacity was all that Faraday required to prove that electrostatic force acted from particle to particle, he knew that his more orthodox colleagues would demand a more direct and conclusive proof. This he set out to provide. Once again, the model was provided by electrochemistry. He had been struck early in his electrochemical researches by a fact that was difficult to reconcile with the theory of electrical forces acting at a distance upon the molecular fragments of a solution. Why, if this were the case, did not the fragments pile up on the sides of the electrodes facing one another? That the decomposing substances plated out evenly on the electrodes was evidence to Faraday that the action was from particle to particle. In order for such even deposition to occur, the action had to be in curved lines and this is what one would expect, given the volume of the particles themselves. Only that line of particles running directly from the center of one electrode to the center of the other would be straight; all the rest would be forced into curves by the volume effect of the molecules. Ever since 1833, Faraday had used this concept to prove that when action was along curved lines, it necessarily meant that the force was intermolecular rather than acting at a distance. In order to prove, then, that electrostatic induction took place from particle to particle he felt it necessary only to show that the line of electrostatic force was a curved, not a straight, line. This he did very simply by charging a rod of shellac and placing on top of it a brass ball. Metals will not conduct the electrostatic line of force, so if electrostatic induction is truly action at a distance, there should be a well-defined "shadow" cast by the ball in which there will be no inductive effect. When the "shadow" was explored with an electrometer, however, induction did take place. The electrostatic line of force could bend around the ball and this, Faraday insisted, was *prima facie* evidence that the electrostatic force was transmitted from particle to particle.

By 1838, the electrostatic line of force had become,

for Faraday, the ultimate and basic reality in all electrical phenomena. With it he was now able to create a unified theory of electrical action which was much simpler and neater than the rival two-fluid theory.

Faraday took the setting up of the electrostatic line of force as the first step in *all* electrical action. This first imposition of intermolecular strain Faraday called the *electrotonic state;* it bears a striking resemblance to what Maxwell later christened the *displacement current*. What happens next depends upon the materials upon which this strain is imposed. Metals cannot bear much strain, so almost immediately upon being set up, the strain collapses, only to be reimposed instantaneously. This build-up and breakdown of the electrotonic state sends a wave of force down the wire. It is this wave which constitutes the electric current. If the material upon which the strain is imposed is an electrolyte, chemical decomposition takes place when the strain is sufficiently strong. Again, the current is exactly the same as in a wire, but it is accompanied by the passage of the elements of the decomposing substance in opposite directions. Thus, electrolytes require somewhat more applied force, or, to put it another way, they are poorer conductors than the metals.

Finally, there are those substances which can take a great amount of strain. The electrotonic state can be built up to a considerable magnitude before the intermolecular forces give way. When this happens, there is a spark, or dielectrical breakdown as when a hole is punched by electrical discharge through a thin sheet of glass. Faraday's theory here beautifully explained an effect that had long puzzled adherents of the two-fluid theory. Lightning always followed the line of least resistance and it appeared as though something went ahead of the lightning bolt to mark out its path. The electrotonic state explained this perfectly for, in effect, it did mark out the path. When the weakest particle in this chain gave way, the lightning passed, following what was indeed the path of no resistance since the chain was broken.

This, then, was Faraday's electrostatic field theory with which he constructed his theory of electrical action. Before we pass on to his magnetic field theory, certain aspects of this early theory should be reiterated and underlined.

We have already noted that the electrostatic line of force was a line of intermolecular strain. This meant that the particles of this line were *polarized*. That is, each particle is "stretched" along the line; this distortion of the force pattern gives rise to the plus and minus ends of the line of the force and, by extension, of the particle itself. This is why the quantity of plus always equals the minus, since both are merely the results of the same distorting force. The electrostatic line of force, therefore, is always a chain of polar particles. This line always has two ends at which the "charges" appear. We shall see that the magnetic line of force is quite a different thing.

What will be preserved in Faraday's later theory is the fundamental tenet of classical field theory, namely the location of the energy of electrical and magnetic action in the *medium* between charged or magnetic poles, and not in the charges or poles themselves.

It was to be some years, however, before Faraday

was able to turn to magnetic phenomena. He had always suffered from poor memory and giddiness. The seven years of concentrated thought and effort which culminated in his great papers of 1838–9 were more than his mind could take. He had a breakdown and was forced to rest. Not until 1845 could he once more focus his mental energies sufficiently to begin to think abstractly once more. From the ensuing decade of effort emerged Faraday's magnetic field theory.

CHAPTER

IV

THE MAGNETIC
LINE OF FORCE

FARADAY'S theory of electricity made a singularly slight impression on his contemporaries. There were a few polite queries from physicists, but by and large the reaction was one of cool disdain. After all, what did a mere chemist really know about the arcana of physics? And how could a mathematical infant like Faraday have the temerity to suggest that the beautiful analytical theory of Coulomb and Poisson might be wrong? It was true, of course, that Faraday's theory had led him to certain discoveries of fact, such as specific inductive capacity, that were not to be expected on orthodox theory, but that could be put down to Faraday's genius as an experimentalist. Comforting themselves with thoughts like these, most European physicists could then return to the contemplation of their integrals, satisfied that the Newtonian way was still the best.

In a situation like this, one must count on the young, for whom the old theories also represent the older generation. In the vigor of youth, it is always good to let fly occasionally at one's elders. So it was here. A young Scot, William Thomson, later Lord Kelvin, thought Faraday's ideas might be worth more than the curt dismissal they had received. With some excitement he discovered that the electrostatic line of force and the electrotonic state were both amenable to mathematical treatment. Perhaps thus clothed they would get the attention Thomson now felt certain they deserved. Curiously enough it was on the experimental side that Thomson thought Faraday's ideas were a bit weak. According to Thomson's analysis, there ought to be some observable facts which Faraday had not noticed. In particular, the electrotonic state as a condition of intermolecular strain should be as easily detectable as a mechanical strain. The standard method in 1845 of detecting such strains if they could be imposed upon a transparent substance was to pass a ray of plane polarized light through the substance while the strain was present. The result was a depolarization of the ray, the extent of the depolarization depending roughly on the degree of strain. So, Thomson suggested that Faraday place a piece of glass between the electrodes of an electrostatic generator and pass a beam of plane polarized light through it to see if the electrotonic state would reveal itself. What Thomson did not know was that Faraday had been trying to detect the electrotonic state in this fashion for twenty years, but to no avail. Nevertheless, Thomson's letter did serve to stimulate Faraday to try once more. This was the beginning of another extended series of experimental researches which extended Faraday's concept of the line of force to include magnetic as well as electrostatic phenomena.

Faraday's first attempts at detecting the electrotonic state were as fruitless as they had been in the 1820's. Yet, with Thomson to back him up, he persevered, thoroughly convinced that both he and Thomson could not be wrong. Every possible change was rung on the experimental conditions. A static field was imposed across a piece of glass; nothing happened. The field was rapidly changed; nothing happened. No matter what Faraday did, nothing happened. Perhaps the fault lay in the magnitude of the forces involved. After all, although quite high voltages could be built up by an electrostatic generator, the electrical forces of attraction and repulsion involved were comparatively weak when compared, say, to an electromagnet. Only small pieces of chaff or cloth would cling to an electrode, whereas the Royal Institution's great electromagnet could lift well over 500 pounds. Moreover, the magnetic forces were exerted along curved lines, and this meant to Faraday that they were the result of intermolecular strain. So, Faraday reasoned, if there were a detectable state of intermolecular strain, it should become manifest under the influence of the full power of the electromagnet. This would not be the same as the detection of the electrotonic state, but the analogy between the electrotonic and magnetic strain was close enough so that detection of the one would give some support to the hypothetical existence of the other.

Using a piece of heavy borate of lead glass, which he had made years ago in some optical experiments, he began his search. This time he was successful. When the heavy glass was placed in a strong magnetic field and a ray of plane polarized light was passed through it in the direction of the magnetic curves of force, there was an effect! The plane of polarization was rotated, the degree of rotation depending both upon the strength of the magnet and the length of the path of the light through the glass. Faraday was triumphant and immediately announced his discovery as further proof of the correctness of his theory.

There was also, however, a nagging doubt that was born in this experiment which eventually led to a reversal of Faraday's ideas. The one thing wrong in the experiment was that the intermolecular strain detected by the polarized light did not act the way ordinary intermolecular strains should. If one took a hollow glass prism filled with turpentine, a substance which rotated the plane of polarized light without the imposition of an external force, and passed a ray of polarized light through it, the rotation was always in the *same* direction in relation to the observer. Thus if the observer's eye were at the left of the page, and the polarized light passed from the right of the page, through the turpentine to the eye, the rotation would be to the right, as far as the observer was concerned. Now, if the observer and the source of polarized light exchange places, the observer will still see a rotation to the right. This is not what happens if a piece of glass is placed in a strong magnetic field. Here the rotation is *absolute*, depending only upon the direction of the magnetic line of force. Thus, if the observer is at the left and the rotation is to the right, when he exchanges positions with the source of light, the rotation will appear to be to the *left*. Only if the polarity of the electromagnet is reversed will the rotation change as in the case of the turpentine. Faraday noted this peculiar effect at the time. Although still reconcilable with the theory of intermolecular strain, it did introduce a complicating factor which was later to assume considerable importance. This factor should be noted here for it contributed to Faraday's increasing attention to the line of magnetic force.

An electrostatic line of force, it will be remembered, was a line of particles undergoing longitudinal strain. The "poles" of this line were the opposite ends of the line of particles where the strain was imposed. The particles, themselves, were polarized, i.e., deformed along the direction of the electrostatic line of force. The polarity here was *in* the particles themselves, and the line of force was nothing but the chain of particles. This was not the case with the magnetic line of force. That the polarity was *not* simply in the particles was shown by the absolute rotation of the polarized light. There seemed to be an interaction between the magnetic line of force and the particles of the heavy glass. Thus, in this case, the line of force was *not* the line of strained particles. It was something independent of them with which they reacted. The independent reality of the line of force was to be the key to Faraday's magnetic field theory.

In 1845, Faraday could only pause briefly over his peculiar discovery. He was now out after bigger game. If light was affected by the magnetic lines of force, then all matter must show some reaction. This had long been believed and Coulomb had gone so far as to announce that all bodies were magnetic like iron, only much more feebly so. Faraday took a somewhat wider view of the problem. He was convinced that so fundamental a force as the magnetic must have some effect on all matter, but he was not convinced that it had to be the same effect as iron exhibited. Thus he was looking for the slightest quiver of a body suspended in a strong magnetic field. This is a significant point for it underlines a fact in the history of science worth marking. People have a tendency to see only what they expect to see and to ignore or forget the unexpected. In fact, the effect Faraday was seeking had been observed at least three times before 1845. But, because it didn't fit into the current theory of magnetism, it simply passed unnoticed. The effect was the setting of the piece of glass across, rather than along, the magnetic lines of force. The glass also was repelled by *both* poles of the magnet. With great excitement Faraday tried everything he could lay his hands on and discovered, to his joy, that no substance was indifferent to the magnet. Everything either acted like iron, setting along the magnetic curves, or like the glass, setting across them. The first class of substances Faraday called *paramagnetics;* the latter, *diamagnetics.* The list of fundamental properties of matter had been extended by one. Once again a startled Europe awoke to find that Faraday's peculiar theory of the action of forces had led him to a fundamental discovery that had escaped his more orthodox colleagues.

The next problem, of course, was to explain what diamagnetism was. Essentially three attempts had been made before Faraday enunciated his own theory. All (including Faraday's) were deficient, but from Faraday's struggle to clarify his own ideas and refute the theories presented by others emerged a new concept of the line of force.

The first account could hardly be called a theory at all since it pretended only to sum up the facts. Yet, the "facts" here were particularly hard to get at and troubled Faraday for some time. It was Hans Christian Oersted who brought this view to the fore: The difference between a paramagnetic and a diamagnetic substance, Oersted suggested, was that a magnetic pole induced a pole of the opposite kind in a paramagnetic and one of the same kind in a diamagnetic. This accounted for the repulsion of a diamagnetic in a magnetic field. Why this difference of magnetic induction in paramagnetic and diamagnetic bodies should exist was a question Oersted left unanswered. He was content merely to describe the difference.

There was one consequence of Oersted's theory of para- and diamagnetic induction that should be detectable but which was singularly difficult to find in the laboratory. A diamagnetic ought to have poles when subjected to magnetic induction, just as a paramagnetic had poles. The only difference should be in the location of these poles. Thus, a bar of bismuth in a magnetic field should merely exhibit a reverse polarity from a bar of iron. Because of the weakness of the diamagnetic force, this polarity was extremely difficult to detect. Faraday performed some simple, but ingenious, experiments in a futile attempt to make it manifest.

The second theory of diamagnetic action was that put forward by Edmond Becquerel. Becquerel's father had worked with Ampère, and Becquerel felt an almost personal interest in defending the theory of magnetism of his famous countryman. He rejected out of hand Faraday's hesitant suggestion that diamagnetism might be something different from ordinary magnetism. Instead he insisted that diamagnetism could be perfectly well explained by assuming that all bodies were magnetic but endowed with different magnetic powers. The repulsion of a substance in a magnetic field could then be understood by an appeal to Archimedes' principle of hydrostatics. The less powerful magnetic substance was displaced by the more powerful. Thus, just as air released under water rose even though it had weight, a bar of bismuth was repelled because the surrounding air was more magnetic than it and pushed it out of the magnetic field. The discovery of the magnetic character of oxygen, unsuspected before Becquerel's theory, appeared to offer strong support for this view.

There were, however, two difficulties. There was still the question of diamagnetic polarity, which was now intensified since the poles to be detected were of an opposite nature according to whether one leaned toward Oersted's or Becquerel's theory. More serious was a theoretical consequence that appeared to be a necessary conclusion and which strained belief a bit too much. In a vacuum, a bar of bismuth was still repelled by a magnetic field. This, said Becquerel, was because the ether was a more magnetic substance than the bismuth. But, according to Ampère, magnetism itself was nothing else but the circulation of electric currents around and through the molecules of the magnetic substance. In a vacuum, there is no magnetic substance, as such, so these electrical currents simply went around in circles by themselves. In the 1860's, an

atomic theory involving circular currents of ether was
to enjoy a brief popularity, but in the 1850's very few
could take such a model seriously.

The third theory was by far the most important. It
was put forward by Wilhelm Weber, who had devoted
his entire scientific life to the study of magnetism. Like
Becquerel, he was convinced that Ampère's model was
the best yet devised but, like Oersted, he was also
certain that diamagnetism was a question of reverse
polarity from paramagnetism. Thus, for Weber, the
detection of this polarity was of primary importance.
Direct methods had already failed since the strength of
the primary magnet needed to produce a diamagnetic
effect masked any feeble manifestations of diamagnetic
polarity. So Weber attacked the problem in an indirect
but seemingly impeccable fashion (see Figure 3).

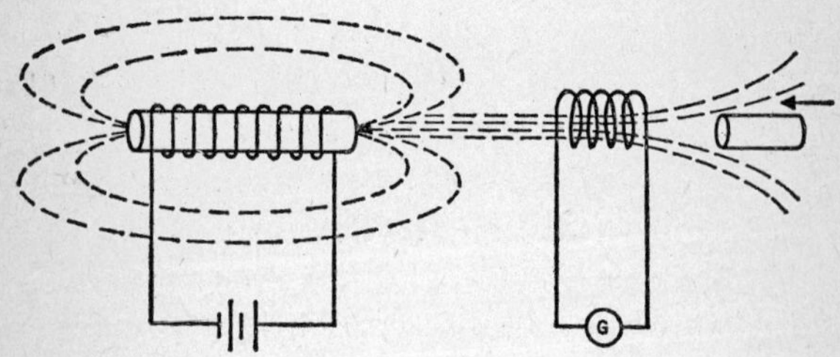

Figure 3

Weber used a cylindrical helix with a soft iron core
as the primary source of magnetism. A short distance
from the end of this bar was a wooden cylinder, coaxial
with the primary helix, around which a coil was wound.
Each end of this coil was connected to the termini of a

galvanometer. The experiment was a simple one. If an iron bar were moved into the wooden cylinder, ordinary magnetic polarity would be induced in it. Its motion, then, would lead to the induction of a current in the detector coil. The *direction* of this current would be indicated by the throw of the galvanometer needle, either to the left or to the right. This direction of throw would then be an accurate indicator of the kind of polarity induced in whatever substances were used. Thus, all bodies which caused the needle of the galvanometer to move in the same direction as the iron had the same polarity. If the needle moved in the opposite direction, then the substance must have received the opposite polarity.

When bismuth was substituted for the iron, the results were all that Weber could have wished. The needle moved in exactly the contrary direction to that observed when the iron was the test substance. Hence, Weber concluded, diamagnets are polar, and their polarity *is* opposite to that of paramagnetics.

Having experimentally proven this fact, Weber set himself to reconciling this effect with Ampère's theory of magnetism. This was no easy task and the weakest part in Weber's theory was here revealed. In order to preserve Ampère's theory, a number of *ad hoc* hypotheses had to be introduced.

According to Ampère, the molecules of magnetic substances were magnetic because of the electrical currents around them. The process of magnetic induction involved the alignment of these molecules so that their currents were all going in the same direction as those in the inducing magnet. Paramagnetic polarity, therefore, was the result of this alignment. Diamagnetic polarity must, by analogy, be the alignment of circular electrical currents which revolved in directions *opposite* to those in the inducing magnet. Now, it was well known that when one current induced another current, the induced current was in the opposite direction from the inducing one. It remained for Weber to adapt this fact to molecular currents.

When a current was induced by either another current or a magnet, the induction took place only when some kind of change in the magnetic field occurred. Thus, a current was induced *only* when the primary current was starting up or stopping, or when the magnet was in motion relative to the wire. Diamagnetism, however, was a steady effect; the bismuth bar remained repelled from the poles of an electromagnet even when the magnetic field did not change. So, Weber had to invent some molecular properties to preserve his induced molecular currents after the inducing force disappeared. Diamagnetic molecules, Weber assumed, must have little channels in them through which the electrical fluids could move without resistance. Once induced by a magnet, therefore, they would continue to circulate even in a steady magnetic field. Why they should cease when the magnet was removed, Weber left unexplained. And why the whole diamagnetic body didn't simply turn around after induction and act like a paramagnet also escaped him.

Throughout the late 1840's and early 1850's, Faraday carefully examined each of the experiments and theories that were published in the scientific press. From the very beginning, he realized that the question of polarity was the central one. Ever since his experiment with the heavy glass, there had been two contending lines of thought in Faraday's mind. On the one hand, he felt he had finally detected the state of intermolecular strain which had been central to his theoretical views for decades. On the other hand, the absolute nature of the rotation of the plane of polarized light seemed to reduce the emphasis on the particles and place it on the line of force. The crucial point was that of polarity. Because of these contending ideas, Faraday was able to entertain the thought that diamagnetic action did not involve a true polarity—a thought quite inconceivable to Oersted, Becquerel or Weber. So, for seven years, Faraday patiently sought the polarity that the one line of his thought suggested, while becoming increasingly convinced of the importance of the other line.

One simple test of diamagnetic polarity immediately suggested itself to Faraday. The patterns formed by iron filings sprinkled on cardboard placed over a magnet clearly revealed the polarity of the filings as they aligned themselves along the line of force. Should not the same thing happen with bismuth filings? The fact that a reversed polarity was induced in the bismuth particles should have no effect on their mutual relations. When the experiment was tried, as Faraday noted in his laboratory *Diary,* "There was no signs of arrangement among the particles inside or outside of the line—no signs of *an attraction* or *repulsion amongst themselves.*"

It might be argued that the diamagnetic force was so feeble that even bismuth filings were too large to be moved by it. To test this, Faraday heated some bismuth until it melted. He then permitted it to cool slowly in an intense magnetic field. Surely if there were diamagnetic polarity would it not be revealed by the regularity of the bismuth crystals aligned by the magnetic force? When the solidified bismuth was carefully broken, no crystalline order whatsoever could be discerned. If there was a diamagnetic polarity, it was singularly elusive.

Just at the point where Faraday's experiments had convinced him that diamagnetic polarity did not exist, Weber's paper detailing his experimental detection of this polarity appeared. Faraday was skeptical, not of Weber's experimental results, but of the conclusion he had drawn from them. Could another interpretation be given to the experiment which avoided the hypothesis of diamagnetic polarity? Faraday did not have to search far for an alternative explanation. Was not the effect created by the bismuth cylinder the result of induced electrical currents in the bismuth, rather than induced diamagnetic polarity? This could be rather simply decided.

Faraday took Weber's experimental setup, but instead of using a bismuth cylinder, he had a copper cylinder cut up into discs. Eddy currents could then be formed in these discs and it would be these currents, rather than an induced diamagnetism, which would affect the detector helix. When the experiment was tried, a strong effect was noted, and the swing of the galvanometer was in a direction contrary to that produced by the iron. To show that the effect was really due to the electrical currents, a glass cylinder was filled with powdered copper. This would prevent free circulation of electrical eddy currents, but should have no effect on any diamagnetic polarity. No effect was produced. Thus Faraday could explain Weber's results without introducing any diamagnetic polarity.

When the iron bar was used, magnetism was induced, and when this magnet passed through the detector helix a current was produced. The effect of the eddy currents was here masked by the magnetic induction. When the bismuth cylinder was used, electrical currents concentric to the axis were induced by the magnetic field of the primary magnetic. These currents, in turn, produced a magnetic field which, cutting the wires of the detector helix, caused a current in this circuit opposite in direction to that produced by the iron. Q.E.D. Never again was Faraday to take diamagnetic polarity seriously.

If the explanation of diamagnetism was not to be found in the forces of the particles of diamagnetics, there was only one other place to look for it. This was in the magnetic line of force itself. By 1850, Faraday's attention was absorbed by these lines and he worked feverishly to lay bare the laws that governed the interaction of lines of force with ponderable matter.

As early as 1845, Faraday had perceived one fact that had contributed to his suspicions of diamagnetic polarity. This was that diamagnetics did not seem to be influenced by magnetic poles as such. Their motions were dictated by the concentration of the lines of force; diamagnetics tended to move from areas of strong magnetism to areas of weak magnetism. This fact led him to suggest an idea that was later to become central.

"I cannot resist," he wrote, "throwing forth another view of these phenomena which may possibly be the true one. The lines of magnetic force may perhaps be assumed as in some degree resembling the rays of light, heat, etc., and may find difficulty in passing through bodies, and so be affected by them, as light is affected." The difference between a paramagnetic and a diamagnetic substance might simply be the result of their respective conducting power for the magnetic line of force. It was an idea worth pursuing and Faraday almost immediately was able to apply it to an experimental situation. It was Weber's old experiment which could now be interpreted in terms of conductibility of substances for the lines of force.

Iron and paramagnetics conduct the lines of force very well. Thus in a magnetic field the lines of force tend to converge on the iron, and the field is intensified within the iron itself. Diamagnetics are poor conductors of the lines of force; therefore, the lines tend to spread out around a diamagnetic, as they are conducted more easily by the medium than by the diamagnetic. Weber's results follow simply from these facts. When the iron is used, the lines of force concentrate upon it, thus cutting the test helix and inducing an electrical current in it. When the bismuth cylinder is substituted, the lines of force are pushed away from it, thus cutting the helix and generating a current in the opposite direction. No polarity of any kind is required and we can even dispense with the eddy currents.

Note that here a single experiment has *three* different interpretations and was, in fact, used to prove three quite different things. For those who like to insist that science is merely a collection of facts, it is worth pondering. To those who like to think of science as a creative, imaginative enterprise, controlled by the acid test of experiment, it may offer some support. It is not enough to know the facts; we must also make some sense of them.

The success of Faraday's new idea in explaining Weber's experiment led him to undertake the more general and difficult task of applying it to all magnetic phenomena. The result was rather spectacular, for everything fell neatly into place. First, induced magnetic polarity was found to fit neatly into the concept of conductibility of lines of force. If an iron sphere were placed between the two poles of a magnet (see Figure 4) the lines of force would converge upon it. Thus, at each end, there would be a convergence of the lines of force and such a convergence could be equated with polarity. Unless the sphere were *exactly* midway between the two magnetic poles (i.e., so that the line ab in the figure passes through the center of the sphere), the concentration on one side would be stronger than on the other. Paramagnetics move into areas of stronger magnetic force; hence the sphere will be "attracted" to one pole or another. Note that action at a distance has disappeared here. The sphere is not "attracted" to the distant pole, but is reacting to the pattern of the lines of force where it, the sphere, is.

The situation of a diamagnetic sphere in similar circumstances was even more instructive. That diamagnetic polarity did not exist could now be seen at a glance (see Figure 5), for if a pole were defined as a point of converging lines of force, it was clear that no such point existed. It also became perfectly clear why a diamagnetic in a uniform magnetic field was content to rest wherever it was placed. A diamagnetic sought to place itself at the point of weakest magnetic action; this

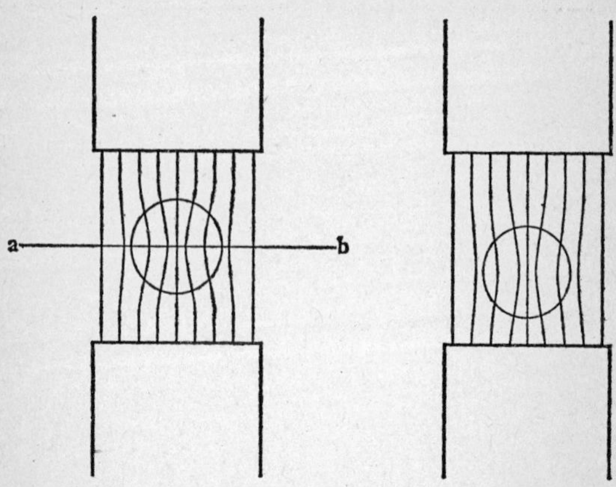

Figure 4

place was always its own center of gravity and so there it stayed. There was a certain triumph in Faraday's voice when he presented this fact, as though he were saying to Weber and Becquerel, "Where are your poles now?"

Finally, Faraday's view harmonized nicely with the fact of differential magnetic action upon which Becquerel had placed so much emphasis. If a diamagnetic were placed in a medium more diamagnetic than it, then the lines of force *would* converge upon it and it would act like a paramagnetic substance. Similarly, paramagnetic substances in media more paramagnetic than they would act like diamagnetics. One did not have to call upon hypothetical electrical currents going this way around the molecules of certain substances and that way around the molecules of others. One

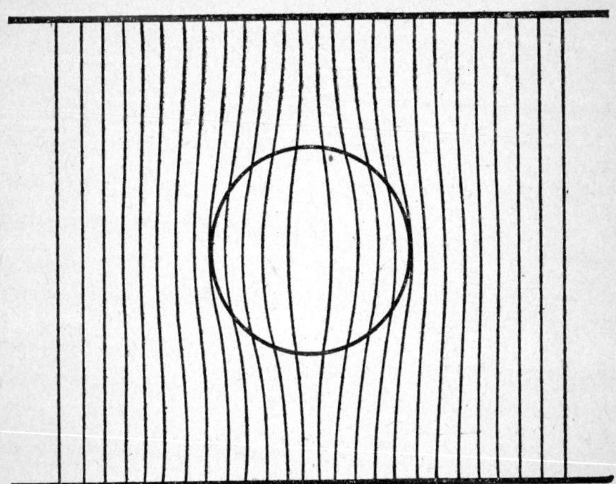

Figure 5

merely had to pay attention to the lines of force and all magnetic phenomena could be predicted.

The ease with which Faraday's concept of the lines of force permitted him to bring order into the confusion of diamagnetic phenomena stimulated him to look more deeply into their nature. Since they now seemed to be the primary aspect of all magnetic phenomena, it would be a good thing to make sure that they really existed. Here is another example of Faraday's subtlety, for he saw that it was possible that the "line of force" so easily detected by iron filings or a compass needle might, in fact, be the result of the presence of the filings or the needle, in which case all his reasonings would have to be revised. The only method of detection which seemed to be above reproach was that of a wire connected to a galvanometer. When the wire and the magnet were mutually at rest, there was no sign of the lines of force. Move the wire, however, and a current was generated indicating that lines of force had been there and had been cut. It seemed hardly likely, Faraday felt, that the mere motion of the wire called the lines of force of the magnet into being, and this method for their detection he considered free from objection.

The use of a moving wire also permitted him to estimate the relative number of lines of force that existed in a given space. The current (or better, the electromotive force) in the detecting circuit was directly proportional to the number of lines of force cut by the circuit. This law was of the greatest importance as he now turned to an intensive study of the lines of force and their relationship to the magnet.

His use of the lines of force to illustrate what "polarity" was when a paramagnetic had been placed in a uniform magnetic field had raised certain doubts in Faraday's mind about the "poles" of a permanent magnet. By analogy with the electrostatic line of force, the poles of a permanent magnet should be the places where the magnetic line of force commenced and terminated. There were many suspicious differences in the analogy that deserved notice. In the first place, a body could be given a single charge; the electrostatic line then passed through the air to another anchoring place. These were clearly termini. A single magnetic pole could not be created and the magnetic line of force did not pass to any other body; it always returned to the body from which it had sprung. The "poles," then, must both be in the magnet. But this made little sense to Faraday. Why, in a homogeneous bar of iron, should two arbitrary points be the places from which the line of force emanated? Did it not make more sense simply to think of the lines of force as passing through the magnet, and the "poles" simply as the points of maximum concentration of the lines of force?

It was one thing to see the lines of force in one's mind's eye as continuous curves; it was another to prove this to be the case. Again, Faraday's experimental setup was simplicity itself. First he repeated an experiment he had performed in 1832. A copper disc was fixed to one end of a cylindrical bar magnet. Leads were then fixed to the center and rim of the disc, and passed to a galvanometer. When the magnet and disc were rotated around the axis of the magnet, a current was detected in the disc. When the disc was separated from the magnet and held still while the magnet was rotated, there was no current. This proved to Faraday's satisfaction that the lines of force did not move with the magnet. The lines of force were not individual lines of strain attached to or generated by specific parts of the magnet. They were general strains produced by the magnet as a whole. Magnets, as an eminent American physicist has since put it, do not grow hairs.

Now Faraday set about detecting the lines of force inside a magnet. He took two square magnets and placed them close together with similar poles at each end. A copper collar was placed at the middle and a sliding lead ran from this collar to a galvanometer. The other terminus of the galvanometer was connected to a wire which ran axially between the two magnets and then up to press against the underside of the copper collar. Each of these three elements could be rotated independently of the others. If the top wire was rotated it would intersect all the external lines of force of the magnet, and the current thus generated could be measured by the galvanometer. Similarly, the axial wire could be rotated on its axis and the part rubbing against the collar would then intersect all the lines of force "inside" this artificial compound magnet. To Faraday's delight, not only was there a current when this wire was rotated, but the current was exactly the same as when the outer wire was rotated. Thus, he concluded, there are lines of force within the magnet and they are exactly equal in amount to the ones outside the magnet. It now seemed obvious that the lines of force were the same ones within and without the magnet. Hence the lines of force were continuous curves. There were no poles; there were not even centers of action. The attack on action at a distance had come full circle. Action at a distance had been associated with central forces since Newton first began to wrestle with the problem of force. Now Faraday had calmly removed the centers. All action depended simply upon the pattern of the lines of force.

There was an elegance and simplicity to Faraday's picture of a magnet that ultimately was to command allegiance to it. Most of Faraday's contemporaries, however, shook their heads sadly to see what age had done to a once powerful mind. What mathematical physicist in his right mind would abandon the logical purity of Ampère's equations for the tangle of lines of force? There was also a difficulty in Faraday's theory that could not be evaded. With the lines of force he could, as we have seen, explain all magnetic phenomena except one. He could not explain what a magnet was! Magnets, he once remarked, are the habitations of lines of force. This, it must be confessed, does not give us a very penetrating insight into the essence of the magnetic state.

Faraday was content to leave this question to one side as he pursued his beloved lines of force into the realm of space itself. Having proven their existence and traced out their continuous character, it was now time to ask just what they were. They were definitely not like the electrostatic lines of force.

> If they exist [he wrote], it is not by a succession of particles, as in the case of static electric induction . . . but by the condition of space free from such material particles. A magnet placed in the middle of the best vacuum we can produce, and whether that vacuum be formed in a space previously occupied by paramagnetic or diamagnetic bodies, acts as well upon a needle as if it were surrounded by air, water or glass; and therefore these lines exist in such a vacuum as well as where there is matter.[1]

This was a rather peculiar situation. The lines of force were strains, but they were not strains of material particles. What then were they? The standard approach to a question such as this in the nineteenth century was to call upon that theoretical workhorse which solved all problems—the ether. Faraday, however, had spent all

[1] Michael Faraday, *Experimental Researches in Electricity,* 3 vols. (London, 1839–55), III, ¶ 3258.

his life getting rid of imponderable, hypothetical fluids and he was not prepared now to found his own theory on such a notion. Furthermore, there was good evidence (Faraday thought) for rejecting the ether. The ether was, itself, assumed to be particulate and if the magnetic line of force was a strain of these particles, there ought to be an ethereal polarity, completely analogous to the electrostatic line of force. No such polarity had ever been detected, and this was enough for Faraday. He was content to state that the magnetic lines of force were lines of strain. If space were empty, so be it: the strains were still there. One is reminded of Schelling's remark that he who required a material substratum for force is no philosopher.

The relation between material bodies and these spatial strains was of the greatest importance for Faraday. To illustrate its significance he once used an analogy which throws considerable light on the relation, in general, between force and matter.

> The magnet . . . may be considered as analogous in its condition to a voltaic battery immersed in water or any other electrolyte. . . . I think the analogy with the voltaic battery so placed, is closer than with any case of *static* electric induction, because in the former instance the physical lines of electric force may be traced both through the battery and its surrounding medium, for they form continuous curves like those I have imagined within and without the magnet.[2]

By comparing the magnet to a voltaic pile immersed in an electrolyte, Faraday was able to make perfectly clear an idea which otherwise might have been difficult

[2] *Ibid.,* ¶ 3276.

to understand. Everyone knew what electrical conduction was and there could be no problem in visualizing the flow of current from one electrode to another through the surrounding electrolyte. Take away the electrolyte and the voltaic cell became an inert container filled with chemicals. Only when the external medium permitted the passage of the electricity, did the cell become a center of electrical force. The same applied to the magnet. "I incline," Faraday wrote, "to consider this outer medium as *essential* to the magnet; that it is that which relates the external polarities to each other by curved lines of power, and that these must be so related as a matter of necessity. Just as in the case of the battery above, there is no line of force, either in or out of the battery, if this relation be cut off by removing or intercepting the conducting medium."

This was the final element in the development of Faraday's field theory. Force could not exist without a medium and the force was to be found *in* the medium, not in the body from which it originated. The magnetic lines of force were propagated in empty space so that one of the properties of this space was its ability to transmit the magnetic strain. It also could transmit another strain, that of gravitational force and Faraday used this property of space to account for both the propagation of light and to direct one final attack against action at a distance.

The electromagnetic theory of light is usually attributed to Maxwell and is often cited as a triumph of classical field theory. Without detracting from Maxwell's contributions to this theory, it should be pointed out that its first enunciation, albeit in a confused and vague way, was given by Faraday in 1846. At this early date then, Faraday had been able to envision the universe as a three-dimensional web of lines of force criss-crossing to infinity. Given these lines, the ether seemed to be an unnecessary hypothesis and it was explicitly to exclude it from physics that he suggested that the lines of force could carry the ray vibrations of light. Thus, light was the vibration of a line of force. When plucked at the source, the line passed this "force" along to its destination. There is a considerable difference between this concept and that of Maxwell, but the similarity is sufficient to justify mention of it here.

In 1857, Faraday tried one last time to convince his contemporaries of the correctness of his field concept and the fundamental error of action at a distance. He had one new weapon to add to the arsenal produced by his own theories and experiments. In the 1840's, there had gradually emerged the recognition of the principle of the conservation of energy, and it was with this principle at the ready that Faraday rode out to do battle. According to him, action at a distance led to the creation of force (or energy) and this was an absurdity. Only in a field theory could this absurdity be avoided.

Suppose that there were but one body, A, in the universe. According to the action-at-a-distance partisans, Faraday insisted, this body had *no* force associated with it. Now suppose that a body B should make its appearance. Instantaneously, body A would attract it and body A would suddenly be endowed with force. Now whence comes this force? It must be created out of nothing, and this violates the principle of the conservation of energy.

How different the picture is when a field concept is used. Here the presence of A sets up gravitational lines of force (or strain) in all space. The appearance of B does not create anything new. B reacts to the strain in space where it, B, is and there is no creation of force. The energy of the system is in the medium; the bodies merely detect these strains and react to them. This action is *never* at a distance but where the body is.

The response of Faraday's colleagues was one of complete disbelief. The reaction of Sir George B. Airy, Astronomer Royal, was typical. Airy rather drily remarked that he could as easily believe in the creation of force in body A as he could in the creation of body B. And that was that!

Not quite. Faraday had sent a copy of his paper on the conservation of force to a young Scot who had evinced an interest in electrical science. He received the following letter in reply:

Dear Sir,

I have to acknowledge receipt of your papers on the Relations of Gold &c. to Light and on the Conservation of Force. Last spring you were so kind as to send me a copy of the latter paper and to ask what I thought of it. That question silenced me at that time, but I have since heard and read various opinions on the subject which render it both easy and right for me to say what I think. And first I pass over some who have never understood the known doctrine of conservation of force and who suppose it to have something to do with the equality of action and reaction. Now first I am sorry that we do not keep our words for distinct things more distinct and speak of the "Conservation of Work or of Energy" as applied to the relations between the amount of "vis viva" and of "tension" in the world; and of the "Duality of Force" as referring to the equality of action and reaction. Energy is the power a thing has of doing work arising either from its own motion or from the "tension" subsisting between it and other things.

Force is the tendency of a body to pass from one place to another and depends upon the amount of change of "tension" which that passage would produce.

Now as far as I know you are the first person in whom the idea of bodies acting at a distance by throwing the surrounding medium into a state of constraint has arisen, as a principle to be actually believed in. We have had streams of hooks and eyes flying around magnets, and even pictures of them so beset, but nothing is clearer than your descriptions of all sources of force keeping up a state of energy in all that surrounds them, which state by its increase or diminution measures the work done by any change in the system. You seem to see the lines of force curving round obstacles and driving plumb at conductors and swerving towards certain directions in crystals, and carrying with them everywhere the same amount of attractive power spread wider or denser as the lines widen or contract.

You have also seen that the great mystery is not how like bodies repel and unlike attract but how like bodies attract (by gravitation). But if you can get over that difficulty, either by making gravity the residual of the two electricities or by simply admitting it, then your lines of force can "weave a web across the sky" and lead the stars in their courses without any necessarily immediate connection with the objects of their attraction.

The lines of Force from the Sun spread out from him and when they come near a planet *curve out from it* so that every planet diverts a number depending on its mass from their course and substitutes a system of its own so as to become something like a comet, *if lines of force were visible.*

. . . Now conceive every one of these lines (which never interfere but proceed from sun & planet to infinity) to have a *pushing* force, instead of a *pulling* one and then sun and planet will be pushed together with a force which comes out as it ought proportional to the product of the masses & the inverse square of the distance.

The difference between this case and that of the dipolar forces is, that instead of each body catching the lines of force from the rest all the lines keep as clear of other bodies as they can and go off to the infinite sphere against which I have supposed them to push.

Here then we have conservation of energy (actual & potential) as every student of dynamics learns, and besides this we have conservation of "lines of force" as to their number and total strength for *every* body always sends out a number proportional to its own mass, and the pushing effect of each is the same.

All that is altered when bodies approach is the *direction* in which these lines push. When the bodies are distant the distribution of lines near each is little disturbed. When they approach, the lines march round from between them, and come to push behind each so that their resultant action is to bring the bodies together with a *resultant* force increasing as they approach.

Now the mode of looking at Nature which belongs to those who can see the lines of force deals very little with "resultant forces" but with a network of lines of action of which these are the final results. So that I for my part cannot realise your dissatisfaction with the law of gravitation provided you conceive it according to your own principles. It may seem very different when stated by the believers in "forces at a distance" but there can be only differences in form and conception not in quantity or mechanical effect between them and those who trace force by its lines. But when we face the great questions about gravitation, Does it require time? Is it polar to the "outside of the universe" or to anything? Has it any reference to electricity? or does it stand on the very foundation of matter—mass or inertia? then we feel the need of tests whether they be comets or nebulae or laboratory experiments or bold questions as to the truth of received opinions.

I have now merely tried to show you why I do not think gravitation a dangerous subject to apply your method to and that it may be possible to throw light on it also by the *embodiment* of the same ideas which are expressed *mathematically* in the functions of Laplace and of Sir W. R. Hamilton in Planetary Theory.

But these are questions relating to the connexion between magneto-electricity and certain mechanical effects which seem to me opening up quite a new road to the establishment of principles in electricity and a possible confirmation of the physical nature of magnetic lines of force. Professor W. Thomson seems to have some new lights on this subject.

Yours sincerely,
JAMES CLERK MAXWELL[3]

---

[3] Institution of Electrical Engineers, London, J. C. Maxwell to M. Faraday (Nov. 9, 1857).

MAXWELL AND CLASSICAL FIELD THEORY

Maxwell's letter to Faraday in 1857 was not the result of a sudden enthusiasm on Maxwell's part for Faraday's ideas. For some time Maxwell had been convinced that Faraday's speculations and unorthodox concepts of electrical and magnetic action offered a potentially fertile field for further exploitation. The difficulty, as Faraday well knew, was that the orthodox physicist refused to take these ideas seriously because of their nonmathematical character. Maxwell's first service to Faraday's system was to remove this objection. On December 10, 1855, and February 11, 1856, he read two papers, "On Faraday's Lines of Force," to the Cambridge Philosophical Society. Thus began the process of turning Faraday's heresies into the orthodoxy of classical field theory.

Maxwell's mind, even at the tender age of twenty-five when he presented these papers, was a very subtle one which often confused and misled his contemporaries. His use of physical analogies and models, in particular, was quite idiosyncratic and there were few who actually saw what Maxwell was getting at. Most of his readers tended to take him literally and the result was the introduction of some extraordinary mechanisms into the pure sanctuary of theoretical physics. As we shall see, Maxwell's papers teem with gushing ideal fluids, gears, idle wheels and such hardware. One French commentator was led to remark upon reading Maxwell's magistral *Treatise on Electricity and Magnetism* that he thought he was to enter the quiet groves of electromagnetic theory only to discover that he had walked into a factory! Maxwell, in his paper on Faraday's lines of force, was careful to warn his readers that his models were not to be taken as physical truth. In fact, their great value lay in permitting one to search for mathematical expressions without committing oneself to a particular physical theory. Maxwell's words on this subject are worth citing, for unless they are understood, Maxwell's approach to field theory will appear singularly clumsy and unsophisticated.

> In order to obtain physical ideas without adopting a physical theory we must make ourselves familiar with the existence of physical analogies. By a physical analogy I mean that partial similarity between the laws of one science and those of another which makes each of them illustrate the other. Thus all the mathematical sciences are founded on relations between physical laws and laws of numbers, so that the aim of exact science is to reduce the

problem of nature to the determination of quantities by operations with numbers.[1]

Maxwell had been led to this position by the striking discovery by William Thomson, later Lord Kelvin, that the mathematical laws of the uniform motion of heat in homogeneous media were identical in form with those of actions at a distance varying inversely as the square of the distance. Nothing could be more *physically* different than gravitation and heat flow, yet mathematically they were identical. This, Maxwell saw, provided the opportunity for considerable progress in electromagnetic theory. If Faraday's ideas could be cast in a conventional mathematical mold, then it might be possible to use the operations of mathematics to deduce further consequences testable by experiment. Thus, the model that Maxwell chose was to be used merely to permit him to put an idea into mathematical form. From that point on, the model lost all physical significance and ultimately disappeared in a differential equation.

This process is most clearly visible in the very first attempt by Maxwell to mathematize Faraday's lines of force. Seizing upon the formal equivalence of the equations of heat flow and action at a distance, Maxwell sought to substitute the flow of an ideal fluid for action at a distance. Faraday's lines of force were to be considered as tubes containing an ideal fluid in which the pressure varied inversely as the distance from the source of the fluid. With this model, Maxwell went on

[1] *The Scientific Papers of James Clerk Maxwell*, W. D. Niven, ed., 2 vols. (New York, Dover Publications, 1966), I, p. 156.

to show how many of the facts of electrostatics and magnetism could be understood mathematically. The key point, of course, was that the action was not at a distance; the energy of the system was in the "tubes of force" (the expression is Maxwell's, not Faraday's) which carried the ideal fluid.

By utilizing a "fluid" Maxwell was also able to achieve an important synthesis between the ideas of Faraday and those of earlier theoreticians. The concepts of potential and of work had been brought into electrical theory earlier in the century by Denis Poisson and George Green. They were an essential part of electrical theories of action at a distance, but Maxwell was now able to show how easily they could be applied to the "tubes" of force. Thus he could convince the most stubborn opponent of Faraday's theory that he was not being asked to accept ideas that contradicted the older theory in any essential theoretical point; he was only being asked to look at electrical and magnetic action from a new point of view which might prove quite useful. This first paper, then, was essentially one of conciliation by which Faraday's heterodoxy could be clothed in orthodox theoretical garb. It was not that Faraday's concept of the line of force was better than the idea of action at a distance; only that it could be counted as mathematically equivalent.

In the second paper, entitled "On Faraday's 'Electrotonic State,' " Maxwell moved to the attack. He now formally confronted Ampère's system with Faraday's. He first pointed out that Ampère's theory was based upon the assumption that the elements of electrical currents acted upon one another along the lines joining them. "This assumption," Maxwell wrote, "is no doubt warranted by the universal consent of men of science in treating of attractive forces considered as due to the mutual action of particles; but at present we are proceeding on a different principle, and searching for the explanation of the phenomena, not in the currents alone, but also in the surrounding medium."[2] To do this, Maxwell turned to Faraday's idea of the electrotonic state.

From the experimental relations discovered by Faraday that connected electromotive force with the cutting of magnetic lines of force, Maxwell was able to derive certain mathematical functions which he christened electrotonic functions. These then permitted the electrotonic state to be defined with mathematical precision.

"We may conceive of the electro-tonic state," he wrote, "at any point of space as a quantity determinate in magnitude and direction, and we may represent the electro-tonic condition of a portion of space by any mechanical system which has at every point some quantity, which may be a velocity, a displacement, or a force, whose direction and magnitude correspond to those of the supposed electro-tonic state. This representation involves no physical theory, it is only a kind of artificial notation. In analytical investigations we make use of the three components of the electro-tonic state, and call them electro-tonic functions. We take the resolved part of the electro-tonic intensity at every point of a closed curve, and find by integration what we

<hr>

[2] *Ibid.,* p. 193.

may call the *entire electro-tonic intensity round the curve.*"[3]

Maxwell's theoretical agnosticism should be noted here and contrasted with Faraday's presentation of the electrotonic state. To Faraday, the electrotonic state was a precise, physical condition of the particles of a body acted upon by the magnetic lines of force. To Maxwell, it was a mathematical function which "involves no physical theory." It can be represented by a variety of physical entities but its essence is its mathematical form. The attraction of this approach lay precisely in its avoidance of physical concepts. Maxwell pointed out this advantage quite explicitly. Ampère's theory, involving two separate electrical fluids, had been developed by Wilhelm Weber in Germany to account physically for the interactions of two electrical currents. In this development, Weber had had to assume that the attraction or repulsion of the moving electrical particles upon one another was dependent upon their velocities. "There are . . . objections," Maxwell wrote, "to making any ultimate forces in nature depend on the velocity of the bodies between which they act. If the forces in nature are to be reduced to forces acting between particles, the principle of the Conservation of Force requires that these forces should be in the line joining the particles and functions of the distance only."[4] Thus Maxwell strongly hinted that the *physical* theory behind Ampère's electrodynamic laws must be abandoned. He was not yet prepared to suggest that his mathematical interpretation of Faraday's the-

[3] *Ibid.*, p. 205.
[4] *Ibid.*, p. 208.

ory was to be substituted for it but he did show that the way pointed out by Faraday might be well worth following for some distance.

Not until 1861 was Maxwell able to return to a further consideration of the properties of the lines of force. By this time, he was becoming more and more convinced by Faraday's arguments in favor of the reality of the lines of force. In a paper "On Physical Lines of Force," he remarked on the impact the patterns of iron filings in a magnetic field had on him.

> The beautiful illustration of the presence of magnetic force afforded by this experiment, naturally tends to make us think of the lines of force as something real, and as indicating something more than the mere resultant of two forces, whose seat of action is at a distance, and which do not exist there at all until a magnet is placed in that part of the field. We are dissatisfied with the explanation founded on the hypothesis of attractive and repellent forces directed towards the magnetic poles, even though we may have satisfied ourselves that the phenomenon is in strict accordance with that hypothesis, and we cannot help thinking that in every place where we find these lines of force, some physical state or action must exist in sufficient energy to produce the actual phenomenon.[5]

The agnosticism had vanished and Maxwell had committed himself to Faraday's views. The problem now was to elucidate them by the application of the mathematical tools of which Maxwell was a master. Before he began, however, Maxwell again pointed out what his method was to be.

> My object in this paper is to clear the way for speculation in this direction, by investigating the mechani-

[5] *Ibid.*, p. 451.

cal results of certain states of tension and motion in a medium, and comparing these with the observed phenomena of magnetism and electricity. By pointing out the mechanical consequences of such hypotheses, I hope to be of some use to those who consider the phenomena as due to the action of a medium, but are in doubt as to the relation of this hypothesis to the experimental laws already established, which have generally been expressed in the language of other hypotheses.[6]

Maxwell was not going to create a physical theory of electricity and magnetism as such. Instead, he was to start with the hypothesis that the medium surrounding electrical currents or magnets must be in certain states of strain or tension. Upon this hypothesis, a mechanical model could be conceived which permitted a mathematical analysis from which certain deductions could be made. These deductions, in turn, could be checked against phenomena. The model was not to be taken as physical reality but only as expressing certain *mechanical* consequences of the assumed state of the medium.

The problem, then, was to fit some kind of a mechanical model to the properties of the medium. Maxwell was no longer content to show that *some* of these properties could be represented mechanically; all must now fit in since the aim of his effort here was to construct a physical theory of the medium in which electromagnetic action took place. There were essentially three properties which a model had to possess if it were to approximate the electromagnetic ether: There must be some way of permitting a linear strain along the lines of force; the mutual lateral repulsion of the lines of force must be accounted for; and the fact that

[6] *Ibid.,* p. 452.

electrical effects seemed always to be at right angles to the magnetic must, somehow, follow from the model.

First, the type of strain had to be established. Were the lines of force lines of tension or compression? From the fact that the lines of force pass from a north pole of one magnet to a south pole of another and that these poles attract one another, Maxwell easily showed that the strain is a tension. The next question was how this tension along the lines of force could be combined with the lateral pressure by which the lines repelled one another. "The explanation which most readily occurs to the mind," Maxwell pointed out, "is that the excess of pressure in the equatorial direction arises from the centrifugal force of vortices or eddies in the medium having their axes in directions parallel to the lines of force.

"This explanation," Maxwell went on, "of the cause of the inequality of pressures at once suggests the means of representing the dipolar character of the line of force. Every vortex is essentially dipolar, the two extremities of its axis being distinguished by the direction of its revolution as observed from those points."[7] The model was really taking shape. In his mind's eye, Maxwell could envision the alignment of these ether vortices, under tension, giving rise to the line of force. Although still to be thought of only as a model, the reality of these vortices was made more likely by the solution of some very difficult hydrodynamic equations. Hermann von Helmholtz showed that a vortex in a frictionless fluid, such as the ether was assumed to be, was astonishingly stable and would continue to exist

[7] *Ibid.*, p. 455.

through eternity. It was even suggested later by Lord Kelvin that the atoms of ponderable matter might be composed of these vortices, thus reducing all physics ultimately to the physics of the ether.

This, however, was in the future. In 1861 Maxwell was intent only upon constructing something upon which he could begin to write what are now his rather well-known equations concerning tensions and centrifugal force. But the physical aspect of the model was not yet complete. The mere rotation of an ether smoke-ring did not account for the generation of an electric current at right angles to the line of force. Nor could the model work as it stood. If the ether vortices were all rotating in a clockwise direction when looked at along the line of force from south to north, then it was impossible for the vortices to press laterally upon one another as demanded by the theory. Two vortices, side by side, have their point of contact moving in opposite directions. If they do touch, then, they must ultimately cause each other to cease rotating. As Maxwell noted,

> I have found great difficulty in conceiving of the existence of vortices in a medium, side by side, revolving in the same direction about parallel axes. The contiguous portions of consecutive vortices must be moving in opposite directions; and it is difficult to understand how the motion of one part of the medium can coexist with, and even produce, an opposite motion of a part in contact with it.[8]

There appeared to be only one way out of this dilemma. Maxwell had to turn from theoretical physics to applied mechanics to find it.

> The only conception which has at all aided me in conceiving of this kind of motion is that of the vortices be-

8 *Ibid.*, p. 468.

ing separated by a layer of particles, revolving each on its own axis in the opposite direction to that of the vortices, so that the contiguous surfaces of the particles and of the vortices have the same motion.

> In mechanism, when two wheels are intended to revolve in the same direction, a wheel is placed between them so as to be in gear with both, and this wheel is called an "idle wheel." The hypothesis about the vortices which I have to suggest is that a layer of particles, acting as idle wheels, is interposed between each vortex and the next, so that each vortex has a tendency to make the neighbouring vortices revolve in the same direction with itself.[9]

The idle wheels were not to be considered as fixed in place, but free to move, although for the most part they would remain in one spot, simply serving to keep the vortices rolling along.

By 1861, Faraday was sinking steadily into senility and there is no evidence that he read Maxwell's paper. It involves no great effort of the imagination to conceive what his reaction would have been had he studied it. The number of purely *ad hoc* elements in Maxwell's model would have horrified him! Faraday had spent his entire career throwing hypothetical entities (like the ether) out of physics. Now, Maxwell was not only bringing the ether back, but endowing it with all kinds of hypothetical properties and motions. And, on top of this, he was insisting upon a whole new kind of particle to make up "idle wheels" to keep the model running properly. Certainly, for Faraday, such a complicated model could bear little resemblance to physical reality. Maxwell undoubtedly felt the same. And yet . . . and yet, it had an amazing ability to account for observed electrical and magnetic phenomena. Using this model

[9] *Ibid.*

as a starting point for his mathematics, Maxwell was able to explain a host of facts. Magnetic attractions and repulsions could be derived by some elementary mathematical operations from the assumed tension and hydrostatic lateral pressure of the rotating vortices. More dramatically, the electrical effects of the disturbance of magnetic lines of force followed so naturally from his model that Maxwell could not repress a certain tone of smugness in his description.

It appears therefore that, according to our hypothesis, an electric current is represented by the transference of the moveable particles interposed between the neighbouring vortices. We may conceive that these particles are very small compared with the size of a vortex, and that the mass of all the particles together is inappreciable compared with that of the vortices, and that a great many vortices, with their surrounding particles, are contained in a single complete molecule of the medium. The particles must be conceived to roll without sliding between the vortices which they separate, and not to touch each other, so that, as long as they remain within the same complete molecule, there is no loss of energy by resistance. When, however, there is a general transference of particles in one direction, they must pass from one molecule to another, and in doing so, may experience resistance, so as to waste electrical energy and generate heat.[10]

It is almost impossible now to doubt that Maxwell was beginning to believe in the physical reality of his model. Certainly no one could blame his readers if they took it as a serious attempt to give a true physical picture of electromagnetism. Nor was this model unsuited for the times. As in his paper on Faraday's lines of force, the net result was to synthesize the two leading

10 *Ibid.*, p. 471.

theories of the day. Faraday's rejection of action at a distance led to a concentration on the nature of the electromagnetic medium; this, in turn, had brought Maxwell to the point of suggesting a mechanism which could easily be appreciated by the more orthodox. Electricity, after all, was reinstated as an imponderable fluid whose only new function was to serve as idle wheels. The ether itself was an old friend serving to transmit the undulations of light. There was no reason why it should not also be called upon to gyrate and be strained to account for magnetism. Even the ether vortices had a certain charm. Ampère had accustomed a whole generation of physicists to think in terms of whirlpools of the electrical fluids. It was really a simplification to substitute the ether. The real advantage, of course, arose from the power of Maxwell's model to explain electricity and magnetism. With severe mathematical logic Maxwell traced out the consequences of his hypothesis. Coulomb's law of electrostatic and magnetic action were deduced from the ethereal strains; Ampère's law of the interaction of two currents also followed from Maxwell's model. The induction of one current by another or by a moving magnet was found to be a natural consequence of the motion of the "idle wheels." Even the Faraday effect—the rotation of the plane of polarized light in a magnetic field—could be shown to be the result of the rotation of the ethereal vortices. Could there really be any doubt that a model which so accurately fit so many observations did not have a real physical existence?

The answer would seem to be yes, for only three years later in his paper "A Dynamical Theory of the Electromagnetic Field" Maxwell quietly abandoned it.

The only physical assumption made now was the existence of an ether. Given this and the mathematical results of his earlier papers, Maxwell constructed a complete theory of the electromagnetic field. It was in this paper that Maxwell laid down the fundamental proposition of classical field theory. The energy of a physical system was not to be found in the material particles of which it was composed but in the ethereal medium surrounding these particles. Maxwell left no room for doubt on this point.

> In speaking of the Energy of the field . . . I wish to be understood literally. All energy is the same as mechanical energy, whether it exists in the form of motion or in that of elasticity, or in any other form. The energy in electromagnetic phenomena is mechanical energy. The only question is, where does it reside? On the old theories it resides in the electrified bodies, conducting circuits, and magnets, in the form of an unknown quality called potential energy, or the power of producing certain effects at a distance. On our theory it resides in the electromagnetic field, in the space surrounding the electrified and magnetic bodies, as well as in those bodies themselves, and is in two different forms, which may be described without hypothesis as magnetic polarization and electric polarization, or, according to a very probable hypothesis, as the motion and the strain of one and the same medium.[11]

This passage really marks the end of our story for it lays out in quite precise terms the definition of classical field theory. The energy of an electromagnetic system could be envisioned as residing in the medium in which the system existed. By 1864, Maxwell had developed mathematical expressions by which the various mani-

[11] *Ibid.*, p. 564.

festations of this energy of the field could be treated with quantitative exactness. In this paper, Maxwell was forced to deal with twenty equations with twenty variables; in his later *Treatise on Electricity and Magnetism*, this number was considerably reduced, but the ideas behind them remained the same.

From 1864 the purely physical aspects of Maxwell's theory became ever more obscured by the mathematical. Note in the passage above the alternative offered by Maxwell to the reader. One could choose either magnetic and electric polarization or motion and strain of a medium. Both led to the same mathematical equations and physical predictions. Late in his life Maxwell wrote at the conclusion of his article on "Ether" for the *Encyclopaedia Britannica*, "Whatever difficulties we may have in forming a consistent idea of the constitution of the aether, there can be no doubt that the interplanetary and interstellar spaces are not empty, but are occupied by a material substance or body, which is certainly the largest, and probably the most uniform body of which we have any knowledge." So it would seem that Maxwell, unlike Faraday, could not dispense with at least one imponderable fluid. It should be noted what he was able to dispense with. The various strain energies of the ether could be *described* by equations. Maxwell never again seriously asked the question of how (in terms of a mechanical analogy) they were produced. He was content to use the equations to account for electromagnetic effects. It was even possible to go beyond what experiment had already revealed. It seemed obvious that the ether which contained the energy of electromagnetism must be the same ether that transmitted the undulations of light. The only question was if there was any close connection between electromagnetic and optical phenomena. From his equations Maxwell found that the velocity of propagation of the magnetic field through space was almost exactly that of light. The connection now seemed clear. "The conception of the propagation of transverse magnetic disturbances to the exclusion of normal ones is distinctly set forth by Professor Faraday in his 'Thoughts on Ray Vibrations.' The electromagnetic theory of light, as proposed by him, is the same in substance as that which I have begun to develop in this paper, except that in 1846 there were no data to calculate the velocity of propagation." The theory was later given complete mathematical expression. Thus physical optics was swallowed up by field theory.

One important consequence followed from Maxwell's equations: The optical spectrum should be only a small region in a much larger range of wave-lengths. There ought to be electromagnetic radiation beyond the optical range that could be created and detected by purely electromagnetic means. When Heinrich Hertz discovered such radiation in 1888, classical field theory appeared to be fully confirmed.

Maxwell's 1864 paper contained all the fundamental ideas that he was to contribute to field theory. In his great *Treatise on Electricity and Magnetism* (1873), the treatment was made more mathematically sophisticated and elegant, but essentially the treatment remained the same. Maxwell's equations did for electromagnetism what Newton's three laws had done two centuries earlier for classical mechanics. And, as had happened with Newton's work, Maxwell's synthesis became the new orthodoxy. No other theory could handle so many phenomena with such ease and accuracy. The number of assumptions necessary to begin the mathematical structure was satisfyingly small, involving only the existence of the ether and the ability of this ether to be strained. As research into the ultimate nature of matter revealed its essentially electrical nature, the hope was even raised that field theory might be the goal towards which men had striven ever since Thales and his fellow Ionian philosophers had first started on the scientific quest—namely, a single theory embracing all physical reality.

There were only a few dark patches to be found. It must always perturb the more discerning to base a theoretical structure entirely upon a substance which eludes all attempts at detection. By the ether's activities it was known, but the ether, *tout seul,* like the phlogiston of eighteenth-century chemistry, could never be looked at, or handled, or weighed, or made manifest. Moreover, Maxwell's equations were esthetically unsatisfactory. They were not symmetrical, as were the equations of classical dynamics. Theoretical mechanics was too beautiful a structure to doubt of its truth; electromagnetism could not quite be brought within this mechanical pale, and this gave rise to certain doubts in the minds of the more astute. These were considered minor flaws, however, in an otherwise impregnable theory. By the end of the nineteenth century, classical field theory was considered to be the most satisfactory general theory in physics. It was not until 1905 that Albert Einstein was to bring it tumbling down with his Special Theory of Relativity. But that is another story.

Selected Bibliography

Cohen, I. Bernard, *Franklin and Newton* (Philadelphia, 1956).

Faraday, Michael, *Faraday's Diary*, Thomas Martin, ed., 7 vols. (London, 1931–1936).

Hesse, Mary, *Forces and Fields* (London, 1961).

Kant, Immanuel, *Metaphysical Foundations of Natural Science*, E. B. Bax, trans. (London, 1883).

Maxwell, James Clerk, *The Scientific Papers of James Clerk Maxwell*, W. D. Niven, ed., 2 vols. (New York, 1966).

Merz, John T., *A History of European Thought in the Nineteenth Century*, 4 vols. (New York, 1965), Vols. 1 and 2.

Oersted, Hans Christian, *Scientific Papers*, 3 vols. (Copenhagen, 1922).

Whyte, Lancelot L., ed., *Roger Joseph Boscovich, S.J.* (London, 1961).

Williams, L. Pearce, *Michael Faraday, A Biography* (London and New York, 1965).

Index